CYBERDETERRENCE
AND CYBERWAR

MARTIN C. LIBICKI

D0980693

Prepared for the United States Air Force
Approved for public release; distribution unlimited

PROJECT AIR FORCE

The research described in this report was sponsored by the United States Air Force under Contract FA7014-06-C-0001. Further information may be obtained from the Strategic Planning Division, Directorate of Plans, Hq USAF.

Library of Congress Cataloging-in-Publication Data

Libicki, Martin C.
    Cyberdeterrence and cyberwar / Martin C. Libicki.
        p. cm.
    Includes bibliographical references.
    ISBN 978-0-8330-4734-2 (pbk. : alk. paper)
        1. Information warfare—United States. 2. Cyberterrorism—United States—
Prevention. 3. Cyberspace—Security measures. 4. Computer networks—Security
measures—United States. 5. Civil defense—United States. I. Title.

U163.L539 2009
355.3'43—dc22

                                                                    2009030055

Published 2009 by the RAND Corporation
1776 Main Street, P.O. Box 2138, Santa Monica, CA 90407-2138
1200 South Hayes Street, Arlington, VA 22202-5050
4570 Fifth Avenue, Suite 600, Pittsburgh, PA 15213-2665
RAND URL: http://www.rand.org/
To order RAND documents or to obtain additional information, contact
Distribution Services: Telephone: (310) 451-7002;
Fax: (310) 451-6915; Email: order@rand.org

Preface

This monograph presents the results of a fiscal year 2008 study, "Defining and Implementing Cyber Command and Cyber Warfare." It discusses the use and limits of power in cyberspace, which has been likened to a medium of potential conflict, much as the air and space domains are. The study was conducted to help clarify and focus attention on the operational realities behind the phrase "fly and fight in cyberspace."

The basic message is simple: Cyberspace is its own medium with its own rules. Cyberattacks, for instance, are enabled not through the generation of force but by the exploitation of the enemy's vulnerabilities. Permanent effects are hard to produce. The medium is fraught with ambiguities about who attacked and why, about what they achieved and whether they can do so again. Something that works today may not work tomorrow (indeed, precisely because it did work today). Thus, deterrence and warfighting tenets established in other media do not necessarily translate reliably into cyberspace. Such tenets must be rethought. This monograph is an attempt to start this rethinking.

The research described in this monograph was sponsored by Lt Gen Robert Elder, Jr., Commander, Eighth Air Force (8AF/CC), and Joint Functional Component Commander for Space and Global Strike, United States Strategic Command. The work was conducted within the Force Modernization and Employment Program of RAND Project AIR FORCE. It should be of interest to the decisionmakers and policy researchers associated with cyberwarfare, as well as to the Air Force planning community.

RAND Project AIR FORCE

RAND Project AIR FORCE (PAF), a division of the RAND Corporation, is the U.S. Air Force's federally funded research and development center for studies and analyses. PAF provides the Air Force with independent analyses of policy alternatives affecting the development, employment, combat readiness, and support of current and future aerospace forces. Research is conducted in four programs: Force Modernization and Employment; Manpower, Personnel, and Training; Resource Management; and Strategy and Doctrine.

Additional information about PAF is available on our Web site:
http://www.rand.org/paf/

Contents

Figures

Tables

Summary

The establishment of the 24th Air Force and U.S. Cyber Command marks the ascent of cyberspace as a military domain. As such, it joins the historic domains of land, sea, air, and space. All this might lead to a belief that the historic constructs of war—force, offense, defense, deterrence—can be applied to cyberspace with little modification. Not so. Instead, cyberspace must be understood in its own terms, and policy decisions being made for these and other new commands must reflect such understanding. Attempts to transfer policy constructs from other forms of warfare will not only fail but also hinder policy and planning.

What follows focuses on the policy dimensions of cyberwar: what it means, what it entails, and whether threats can deter it or defense can mitigate its effects. The Air Force must consider these issues as it creates new capabilities.

Cyberattacks Are Possible Only Because Systems Have Flaws

As long as nations rely on computer networks as a foundation for military and economic power and as long as such computer networks are accessible to the outside, they are at risk. Hackers can steal information, issue phony commands to information systems to cause them to malfunction, and inject phony information to lead men and machines to reach false conclusions and make bad (or no) decisions.

Yet system vulnerabilities do not result from immutable physical laws. They occur because of a gap between theory and practice. In theory, a system should do only what its designers and operators want it to. In practice, it does exactly what its code (and settings) tells it to. The difference exists because systems are complex and growing more so.

In all this lies a saving grace. Errors can be corrected, especially if cyberattacks expose vulnerabilities that need attention. The degree to which and the terms by which computer networks can be accessed from the outside (where almost all adversaries are) can also be specified. There is, in the end, no forced entry in cyberspace. Whoever gets in enters through pathways produced by the system itself.[1] It is only a modest exaggeration to say that organizations are vulnerable to cyberattack only to the extent they want to be. In no other domain of warfare can such a statement be made.

Operational Cyberwar Has an Important Niche Role, but Only That

For operational cyberwar—acting against military targets during a war—to work, its targets have to be accessible and have vulnerabilities. These vulnerabilities have to be exploited in ways the attacker finds useful. It also helps if effects can be monitored.

Certainty in predicting the effects of cyberattacks is undermined by the same complexity that makes cyberattacks possible in the first place. Investigation may reveal that a particular system has a particular vulnerability. Predicting what an attack can do requires knowing how the system and its operators will respond to signs of dysfunction and knowing the behavior of processes and systems associated with the system being attacked. Even then, cyberwar operations neither directly harm individuals nor destroy equipment (albeit with some exceptions). At best, these operations can confuse and frustrate operators of mili-

---

[1]   Distributed denial-of-service attacks, the sort perpetrated against Estonia in 2007, are a partial exception. They clog the entryways to the system, rather than get into it. However, such attacks are, at worst, a minor nuisance to organizations (e.g., the military, electric power producers) that can run without interacting with the public at large.

tary systems, and then only temporarily. Thus, cyberwar can only be a support function for other elements of warfare, for instance, in disarming the enemy.

The salient characteristics of cyberattacks—temporary effects and the way attacks impel countermeasures—suggest that they be used sparingly and precisely. They are better suited to one-shot strikes (e.g., to silence a surface-to-air missile system and allow aircraft to destroy a nuclear facility under construction) than to long campaigns (e.g., to put constant pressure on a nation's capital). Attempting a cyberattack in the hopes that success will facilitate a combat operation may be prudent; betting the operation's success on a particular set of results may not be.

Strategic Cyberwar Is Unlikely to Be Decisive

No one knows how destructive any one strategic cyberwar attack would be. Estimates of the damage from *today's* cyberattacks within the United States range from hundreds of billions of dollars to just a few billion dollars per year.

The higher dollar figures suggest that cyberattacks on enemy civilian infrastructures—strategic cyberwar—may be rationalized as a way to assist military efforts or as a way to coerce the other side to yield to prevent further suffering. But can strategic cyberwar induce political compliance the way, say, strategic airpower would? Airpower tends to succeed when societies are convinced that matters will only get worse. With cyberattacks, the opposite is more likely. As systems are attacked, vulnerabilities are revealed and repaired or routed around. As systems become more hardened, societies become less vulnerable and are likely to become more, rather than less, resistant to further coercion.

Those who would attempt strategic cyberwar also have to worry about escalation to violence, even strategic violence. War termination is also not trivial: With attribution so difficult and with capable third parties abounding (see below), will it be clear when one side has stopped attacking another?

Cyberdeterrence May Not Work as Well as Nuclear Deterrence

The ambiguities of cyberdeterrence contrast starkly with the clarities of nuclear deterrence. In the Cold War nuclear realm, attribution of attack was not a problem; the prospect of battle damage was clear; the 1,000th bomb could be as powerful as the first; counterforce was possible; there were no third parties to worry about; private firms were not expected to defend themselves; any hostile nuclear use crossed an acknowledged threshold; no higher levels of war existed; and both sides always had a lot to lose. Although the threat of retaliation may dissuade cyberattackers, the difficulties and risks suggest the perils of making threats to respond, at least in kind. Indeed, an explicit deterrence posture that encounters a cyberattack with obvious effect but nonobvious source creates a painful dilemma: respond and maybe get it wrong, or refrain and see other deterrence postures lose credibility.

The case for cyberdeterrence generally rests on the assumption that cyberattacks are cheap and that cyberdefense is expensive. If cyberattacks can be conducted with impunity, the attacker has little reason to stop. Besides, nuclear deterrence prevented the outbreak of nuclear conflict during the Cold War. What is there about cyberspace that would prevent a similar posture from working similarly well? Plenty, as it turns out. Questions that simply do not crop up with nuclear or even conventional deterrence matter in cyberspace whenever the target of an attack (the "we") contemplates retaliation.

Will we know who did it? Cyberattacks can be launched from literally anywhere, including cybercafés, open Wi-Fi nodes, and suborned third-party computers. They do not require expensive or rare machinery. They leave next to no unique physical trace. Thus, attribution is often guesswork. True, ironclad attribution is not necessary for deterrence as long as attackers can be persuaded that their actions may provoke retaliation. Yet some proof may be necessary given (1) that the attacker may believe it can shake the retaliator's belief that it got attribution right by doing nothing different ("who, me?") in response to retaliation, (2) that mistaken attribution makes new enemies, and (3) that neutral observers may need to be convinced that retaliation is not aggression.

Can retaliators hold assets at risk? It is possible to understand the target's architecture and test attack software in vivo and still not know how the target will respond under attack. Systems vary by the microsecond. Undiscovered system processes may detect and override errant operations or alert human operators. How long a system malfunctions (and thus how costly the attack is) will depend on how well its administrators understand what went wrong and can respond to the problem. Furthermore, there is no guarantee that attackers in cyberspace will have assets that can be put at risk through cyberspace.

Can they do so repeatedly? It is difficult to imagine an act of cyberretaliation that is prospectively so awesome that no potential attacker would run the risk of being hit (a crucial feature of nuclear retaliation). Repeated application may be necessary but is not necessarily possible. Even successful retaliation may not be convincing if the attacker tells itself it will be less vulnerable the next time around.

Can cyberattacks disarm cyberattackers? In a world of cheap computing, ubiquitous networking, and hackers who could be anywhere, the answer is no.

Will third parties stay out of the way? Cyberattack tools are widely available. If nonstate actors jump into such confrontations, they could complicate attribution or determining whether retaliation made the original attackers back off.

Might retaliation send the wrong message? Most of the critical U.S. infrastructure is private. An explicit deterrence policy may frame cyberattacks as acts of war, which would indemnify infrastructure owners from third-party liability, thereby reducing their incentive to invest in cybersecurity.

Can states set thresholds for response? Unless a state declares that all cyberattacks, no matter how minor, merit retaliation, it needs to define an actionable threshold. Proving that any one attack crossed it, however, may be tricky.

Can escalation be avoided? Even if retaliation is in kind, counterretaliation may not be. A fight that begins in cyberspace may spill over into the real world with grievous consequences.

Responses to Cyberattack Must Weigh Many Factors

In many ways, cyberwar is the manipulation of ambiguity. Not only do successful cyberattacks threaten the credibility of untouched systems (who knows that they have not been corrupted?) but the entire enterprise is beset with ambiguities. Questions arise in cyberwar that have few counterparts in other media.

What was the attacker trying to achieve? Because cyberwar can rarely break things much less take things, the more-obvious motives of war do not apply. If the attacker means to coerce but keep its identity hidden, will the message be clear? If the attack was meant to disarm its target but does so only temporarily, what did the attacker want to accomplish in the interim? Can an attack and its aftermath be used as part of a competitive strategy in commercial or political markets? What role might a cyberattack play in the attacker's master narrative?

What should the target reveal about the attack? Many attacks— corruption attacks, disruption attacks on systems deep within an organization—have effects that are not generally obvious. Revealing what happened is more honest and necessary to justify public retaliation. However, silence might mitigate panic, preserve confidence in systems as they are being fixed, and support nonconfrontational strategies (e.g., private exposure followed by diplomatic threats) or nonpublic retaliatory strategies. Whether and when to name the attacker also deserve thought. Premature revelation can be embarrassing, but if revelation comes long after the attack, the link between retaliation and the original attack may lose credibility. Revelation too far in advance of retaliation gives the attacker time to ward off a retaliatory attack through better defenses, counterthreats, or mobilizing opinion to its side.

How should states respond to freelance attacks? Establishing that a state is protecting the attackers creates yet another hurdle to attribution, but what if the hackers were just not sought with sufficient vigor? It is hard to know whether retaliating against such a state would energize its prosecutorial energies—or backfire.

Should deterrence be extended to allies? Figuring out who actually hit the ally's system and with what effect requires poking into their systems, something they may balk at ("don't you trust us?"). Allies may also have ulterior motives for fingering one particular attacker.

Military Cyberdefense Is Like but Not Equal to Civilian Cyberdefense
Because military networks mostly use the same hardware and software as civilian networks, they have mostly the same vulnerabilities. Their defense resembles nothing so much as the defense of civilian networks—a well-practiced art. But military networks have unique features: real enemies, specific cyberthreats, and many closed systems.

The primary goal for the military is to function as well under cyberattack as it does on a day-to-day basis—after all, performance under military attack is how militaries are measured. Robustness is key, but it goes beyond network security engineering to encompass all measures that permit the broader system (the military itself) to work when its subsystems do not. The military must pay more attention than others do to the failure modes that are likely to be the most damaging or most prevalent.

Because the effects of cyberattack are temporary, the military's first priority in the wake of a major successful attack is to figure out whether a physical attack is coming to take advantage of the systems being crippled. The second priority, assuming the attacker is monitoring such systems before deciding whether to attack, is to make it look as though little damage has been done. The third is to achieve recovery. Everything else (including retaliation) follows later.

Implications for the Air Force

The United States and, by extension, the U.S. Air Force, should not make strategic cyberwar a priority investment area. Strategic cyberwar, by itself, would annoy but not disarm an adversary. Any adversary that merits a strategic cyberwar campaign to be subdued also likely possesses the capability to strike back in ways that may be more than annoying.

Similar caution is necessary when contemplating cyberdeterrence. Attribution, predictable response, the ability to continue attack, and the lack of a counterforce option are all significant barriers. The United States may want to exhaust other approaches first: diplomatic, economic, and prosecutorial.

Operational cyberwar has the potential to contribute to warfare—how much is unknown and, to a large extent, unknowable. Because a devastating cyberattack may facilitate or amplify physical operations and because an operational cyberwar capability is relatively inexpensive, it is worth developing. That noted, success at cyberwar is not only a matter of technique but also requires understanding the adversary's networks in the technical sense and, even more, in the operational sense (how potential adversaries use information to wage war). The Air Force should also recognize that the best cyberattacks have a limited shelf life and should be used sparingly.

Throughout all this, cyberdefense remains the Air Force's most important activity within cyberspace. Although most of what it takes to defend a military network can be learned from what it takes to defend a civilian network, the former differ from the latter in important ways. Thus, the Air Force must think hard as it crafts its cyberdefense goals, architectures, policies, strategies, and operations.

Acknowledgements

RAND work profits enormously from helpful hands and helpful hints. This monograph is no exception, and many individuals deserve heartfelt acknowledgements. First is the RAND team that worked on the overall project. Its members include Richard Mesic, who strongly encouraged this line of inquiry; Robert Anderson; Myron Hura; Lynn Scott; and Lara Schmidt. Donald Stevens, who oversees the group in which the project was conducted, also deserves special thanks. Second are our Air Force sponsors, who oversaw the effort and provided encouragement throughout, notably Lt Gen Robert Elder (Commander Eighth Air Force) and Maj Gen William Lord (Commander 24th Air Force). Third are the many individuals who looked at this document and shared their comments with the author: Greg Rattray; Milt Johnson (Air Force Space Command); and RAND colleagues Paul Davis, James Quinlivan, David Frelinger, Roger Molander, and David Gompert. Fourth are the good folks who read the manuscript in the context of RAND's quality-assurance policy: John Arquilla, Ryan Henry, and Cynthia Cook. Special thanks are also in order for Ricardo Sanchez, who provided invaluable research support; for Jerry Sollinger, for his assistance in communicating the results; and for Catherine Piacente and Karen Suede, who helped prepare the manuscript.

Abbreviations

AT&T	current corporate identity
BDA	battle damage assessment
CNE	computer network exploitation
CNN	Cable News Network
DDOS	distributed denial-of-service
DEC	Digital Equipment Corporation
DNA	deoxyribonucleic acid
DNS	Domain Name Service
DoD	Department of Defense
FBI	Federal Bureau of Investigation
HD DVD	high-definition digital video disc
IBM	International Business Machines
IO	information operations
IPv6	Internet Protocol version 6
MO	modus operandi
NATO	North Atlantic Treaty Organization

NSA	National Security Agency
PLA	Peoples' Liberation Army
RF	radio frequency
RSA	a corporation name
SAM	surface-to-air missile
sysadmin	system administrator
Wi-Fi	trademark

CHAPTER ONE

Introduction

In late April 2007, Estonia moved a Russian war memorial from central Tallinn to a military cemetery, outraging its Russian-speaking population and thereby leading to two days of riots. While this was going on, the country's key Web sites (notably government and bank sites) were flooded by a distributed denial-of-service (DDOS) attack carried out by thousands of hijacked computers (also known as bots).[1]

Although Estonia is small, it is well wired (Skype was launched there), and the country had accustomed itself to being able to conduct business in cyberspace. Furthermore, several hundred thousand Estonians work overseas, and their ability to wire money back is essential to feeding their families. Because some of the relevant rogue packets could be traced back to the Kremlin, many assumed the Russians launched this cyberattack.

In response, Estonia closed its borders to external Internet access (very few of the bots involved were actually inside Estonia). This allowed in-country users, but not external users, to access the sites. Estonia also contracted with routing firms to add redundancy to its external Internet connections.[2] The attacks stopped after a few weeks; what effect Estonia's own efforts had in this are unclear.

---

[1] "Europe: A Cyber-Riot; Estonia and Russia" *The Economist* (London), Vol. 383, No. 8528, May 12, 2007, p. 42.

[2] After the cyberattack, Estonia decided to allow non-Estonian Internet traffic access to its network only through a mirrored site. Visitors now see Estonian information through servers run by Akamai Technologies; see Jason Miller, "Feds Take 'Cyber Pearl Harbor' Seriously," Homeland Security and Defense Business Council, May 29, 2007. Network analysts

Russia denied responsibility but, at the same time, did not welcome investigators seeking to determine what really happened. In January 2008, an Estonian (of ethnic Russian descent) was convicted of perpetrating a least part of the attack.[3] Considerable evidence suggests that he had help from parts of the Russian *mafiya*, which helped organize the hijacked computers for him,[4] but there is still no solid evidence that the Russian state was involved. Since investigations continue—still with no help from Russia—this may not be the last word on who did it.

This incident crystallizes many of the conundrums of cyberdeterrence and cyberwar. First, even today, the identity of the attackers remains unclear. Although rogue packets came from Russia, that does not prove that they originated there; even if the packets originated there, that does not mean the Russian government sent them.[5] The

---

at AT&T (the world's largest Internet service provider) believe that the attacks on Estonia, at 100 Mbps, were small compared to the 4 Gbps attacks they had seen elsewhere.

[3] "Estonia Fines Man for 'Cyberwar,'" *BBC News*, January 25, 2008.

[4] Creating networks of hijacked computers is also sometimes called "herding botnets."

[5] Opinion is still divided on the issue of Russia's responsibility. Stephen Blank, "Web War I: Is Europe's First Information War a New Kind of War?" *Comparative Strategy*, Vol. 27, No. 3, 2008, makes a sophisticated case for Russian culpability. Yet his case is built on (1) links between the Russian government and the rioters, (2) Russia's willingness to bully its neighbors, (3) Russian knowledge of such techniques, and (4) Russia's fear that it could be the victim of information warfare. The paper introduces no new facts about the cyberattacks as such. "Accordingly," the author writes (p. 228), "it seems clear that the computer attacks . . . were sanctioned by high policy and reflected a coordinated strategy devised in advance of the removal of the Bronze Soldier from its original pedestal." Taken at face value, not even that statement is actionable from a deterrence point of view.

Similar accusations of Russian cyberwar against Georgia were made in 2008 (Tom Espiner, "Georgia Accuses Russia of Coordinated Cyberattack," *CNET News*, August 11, 2008b; Siobhan Gorman, "Georgia States Computers Hit By Cyberattack," *Wall Street Journal*, August 12, 2008, p. A9; Russian Business Network, "Georgia CyberWarfare," blog posting, August 9, 2008; Kim Hart, "Longtime Battle Lines Are Recast in Russia and Georgia's Cyberwar," *Washington Post*, August 14, 2008, p. D01; Robert Vamosi, "Kids, not Russian Government, Attacking Georgia's Net, Says Researcher," *CNET News*, August 13, 2008. In Georgia's case, there is also the strong possibility that the Russian Business Network (thought to be criminal) has been controlling access to Georgian servers, linked as they are through Russian networks. Suffice it to say that in mid-August 2008, cyberattacks were not Georgia's topmost worry.

attacks could have been mounted by outraged hackers either in Estonia itself or in Russia, by members of the Russian mafiya, or by overzealous Russian government agents. If members of the last group were the perpetrators, they could have acted either in secret, with the knowledge that their superiors would turn a blind eye, or at the direction of the Russian government. Second, given the unclear nature of the attacker, any retaliation could easily have targeted the wrong source, possibly initiating an escalating series of tit-for-tat attacks. Even if Estonia had had a deterrence policy, it likely would not have forestalled the attack. Such incidents as the one in Estonia and the increasingly bold forays that what may be the China's Peoples' Liberation Army (PLA) has supposedly made into unclassified defense and civilian networks have convinced many that cyberspace has, finally, matured as a medium of conflict and that strategic warfare there has arrived.[6] As a medium, it is automatically strategic in the sense that civilian infrastructures are at no less risk than military ones and can be attacked as easily—indeed more easily, given the great attention militaries tend to pay to security.

The increasing salience of cyberspace and the growing importance of security issues within it are hardly surprising. There are over a billion personal computers, most of which are connected to the Internet.[7] In early 2008, the number of cellular telephone owners worldwide exceeded the population (children included) of nonowners.[8] Every digital cell phone (soon all will be digital) can be a door into cyberspace. Most computer users care little and know less about security. One result is that millions, perhaps tens of millions, of computers today are bots, capable of being controlled by nefarious others their owners have never

---

[6]   Brian Grow, Keith Epstein, and Chi-Chu Tschang, "The New E-spionage Threat," *BusinessWeek,* April 21, 2008, pp. 32–41. See also The White House, Executive Order 13010, *Critical Infrastructure Protection*, and Robert Vamosi, "Cyberattack in Estonia—What It Really Means," *CNET News,* May 29, 2007.

[7]   Siobhan Chapman, "Worldwide PC Numbers to Hit 1 Billion in 2008, Forrester Says," *CIO.com,* June 11, 2007.

[8]   Kirstin Ridley, "Global Mobile Phone Use to Hit Record 3.25 Billion," Reuters, June 27, 2007.

met.[9] Those concerned about security are finding out that the borders between systems are becoming fuzzier and gauzier. Few computers are islands anymore (although most of those that need to be still are). Borders fade further every time users cannot differentiate between what is happening on their own machines and what is happening on others. Suffice it to say that computers are growing ever more complex, and their actions are growing less intuitive. The distinction between data (which are acted on and thus cannot act malevolently) and an application (which acts and thus can act badly) is almost invisible. Modern application software constantly rewrites data and application files even without prompting from users or events. To top this off, systems are growing more important. The decisions they make, which once supplemented human judgment, now often replace it. Anyone who believes that systems are more open to tampering and believes that this tampering has greater consequence has reason to worry despite the growing attention system and software houses are paying to security.

These facts may have strategic ramifications, which echoes earlier arguments that airpower permits nations to win wars without armies (the North Atlantic Treaty Organization's [NATO's] air war over Kosovo suggests they may be getting close).[10] Advocates of cyberspace assert that gaining control of it will achieve information dominance; superiority in every other medium is just a matter of time.[11] Ensuring that the United States (or China, or Russia, or any other nation) can control cyberspace, the argument continues, requires the creation, enlargement, and maintenance of cyberattack and cyberdefense capabilities.

To devise strategies to ensure that no nation can do to the United States what it would like to be able to do to others, continue advocates,

---

[9] We are not counting the process by which software and hardware companies update user machines even as users are only vaguely aware that this is happening.

[10] See Gregory J. Rattray, *Strategic Warfare in Cyberspace*, Cambridge, Mass.: Massachusetts Institute of Technology, 2001, esp. Ch. 4, for a comparison of strategic airpower and strategic cyberwarfare.

[11] For an argument from one Chinese perspective, see Liang Qiao and Wang Xiangsui, *Unrestricted Warfare*, Beijing: PLA Literature and Arts Publishing House, 1999.

the nation should hearken back to the golden age of nuclear strategic thinking. Just as the nuclear era spawned policies of deterrence that, although elaborate, were successful or at least not challenged,[12] today's era needs a doctrine of cyberdeterrence. Indeed, if one could gloss over the technological differences between nuclear weapons and cyberweapons, one might argue that adapting such deterrence policies would suffice for a new medium in the 21st century.

Alas, matters are not so simple; clarity in thinking about cyberdeterrence and cyberwar does require appreciating the unique nature of cyberspace. As with any posture, or course of action, one must ask: What are we trying to achieve, under what circumstances?

Purpose

The need for well-grounded thinking in this medium is no longer academic. The U.S. Air Force has recognized both the potential and the vulnerabilities of cyberspace and has stood up the 24th Air Force to wage war in cyberspace. Significant decisions are being made regarding this new command. The purpose of this monograph is to focus on the policy dimensions of cyberwar: what it means, what it entails, and whether it is possible to deter others from resorting to it. Because cyberwar and cyberdeterrence cannot be understood in isolation, the monograph explores some key aspects of cyberwar to establish a framework for considering cyberdeterrence. It also offers some issues for the Air Force to consider as it creates its new command.

We argue that, because cyberspace is so different a medium, the concepts of deterrence and war may simply lack the logical foundations that they have in the nuclear and conventional realms. Indeed, the

---

[12] The body of literature on the subject is huge. See, for instance, the reference lists from Frank C. Zagare and D. Marc Kolgour, "Deterrence Theory and the Spiral Model Revisited," *Journal of Theoretical Politics*, Vol. 10, No. 1, 1998, pp. 59–87; Mark Irving Lichbach, "Deterrence or Escalation? The Puzzle of Aggregate Studies of Repression and Dissent," *Journal of Conflict Resolution*, Vol. 31, No. 2, 1987, pp. 266–297; and Abram N. Shulsky, *Deterrence Theory and Chinese Behavior*, Santa Monica, Calif.: RAND Corporation, MR-1161-AF, 2000.

range of circumstances under which either or both are worth embarking on could be quite narrow. Ironically, operational cyberwar—cyberattacks to support warfighting—may have far greater purchase than strategic cyberwar, cyberattacks to affect state policy. Correspondingly, the greatest danger to the United States from cyberspace (as well as space) may be operational rather than strategic. If states with powerful militaries come to believe that a sudden cyberattack on the U.S. military could paralyze it long enough for conventional militaries to run roughshod over U.S. interests, the risks they run may endanger us all.

Furthermore, similar principles should also characterize how the United States in general and the Air Force in particular conduct cyberwar. An operational cyberwar capability may well be an effective niche weapon if correctly timed (notably, at the outset of hostilities) and if carefully prepared through diligent and persistent intelligence on the target. Strategic cyberwar campaigns are more problematic and hence merit less emphasis.

Basic Concepts and Monograph Organization

Although the concept of cyberspace is plastic and contentious,[13] our purposes can be served if cyberspace is defined as analogous to the Internet. Cyberspace, as such, can be characterized as an agglomeration of individual computing devices that are networked to one another (e.g., an office local-area network or a corporate wide-area network) and to the outside world. This is not meant to be a comprehensive defi-

---

[13] The Department of Defense (DoD) defines cyberspace as follows:

> A global domain within the information environment consisting of the interdependent network of information technology infrastructures, including the Internet, telecommunications networks, computer systems, and embedded processors and controllers. (Joint Publication 1-02, *DoD Dictionary of Military Terms*, Washington, D.C.: Joint Staff, Joint Doctrine Division, J-7, October 17, 2008.)

Dan Kuehl lays out a cacophony of other definitions in "From Cyberspace to Cyberpower: Defining the Problem," in Franklin Kramer et al., *Cyberpower and National Security*, Washington D.C.: National Defense University Press, 2009, pp. 24–42.

nition; the distinction between a cell phone network and the Internet is becoming harder to make with every day. One can even imagine computers based on DNA rather than electronics.[14] But this definition serves to ground what follows in something that already exists and what most readers are familiar with. Conflict in other media (e.g., over-the-air radio-frequency [RF] transmission) will resemble conflict in cyberspace in some respects but not others. Our intent is to use a narrower definition that permits us to make some generalizations, leaving it to others to extend these generalizations to related media. Chapter Two characterizes cyberspace for the purposes of this discussion.

The concept of deterrence also needs to be defined for our purposes. William Kaufman maintained,

> Deterrence consists of essentially two basic components: first, the expressed intention to defend a certain interest; secondly, the demonstrated capability actually to achieve the defense of the interest in question, or to inflict such a cost on the attacker that, even if he should be able to gain his end, it would not seem worth the effort to him.[15]

Cyberdeterrence can also be discussed in the context of escalation control: a disincentive to escalate or a disincentive to carry out the next attack.

If deterrence is anything that dissuades an attack, it is usually said to have two components: deterrence by denial (the ability to frustrate the attacks) and deterrence by punishment (the threat of retaliation). For purposes of concision, the use of deterrence in this work refers to deterrence by punishment. This is not to deny that defense has no role to play—indeed, the argument here is that it does play the greater role and rightfully so. Our discussion of deterrence (by punishment) asks whether it should be added to defense (deterrence by denial). Also, as

---

[14] Yaakov Benenson, Binyamin Gil, Uri Ben-Dor, Rivka Adar, and Ehud Shapiro, "An Autonomous Molecular Computer for Logical Control of Gene Expression," *Nature,* Vol. 429, No. 6990, May 27, 2004, p. 423.

[15] William Kaufmann, "The Evolution of Deterrence 1945–1958," unpublished RAND research, 1958.

explained below, deterrence by denial and deterrence by punishment are synergistic with one another in some ways. Nevertheless, from this point on, deterrence refers to deterrence by punishment; the rest is defense.

Whether deterrence actually obtains, is, however, an empirical question. Deterrence has to work in the mind of the attacker. Any potential attacker is bound to weigh the effort required to make an attack against the expected benefit of that attack (a function of how likely it is to work and what happens if it does). The point of a deterrence policy is to add another consideration to the attacker's calculus, and that is a function of whether the attacker believes the threat to retaliate will be carried out and the potential damage that will result if and when the retaliation occurs. For reasons explained in Chapter Two, the assumption here is that retaliation is in kind, not because other forms of retaliation are impossible but to better highlight salient aspects of deterrence. Chapter Three discusses cyberdeterrence and explains its profound difference from nuclear deterrence; Chapter Four examines the enemy's motivations for a cyberattack; and Chapter Five considers questions entailed in responding to a cyberattack.

Strategic cyberwar (Chapter Six) is a campaign of cyberattacks one entity carries out on another. It can be unilateral, but this discussion assumes that it is two-sided. This assumption, again, also brings out more aspects of the issue. It is possible to regard strategic cyberwar as what happens when attack, retaliation, and counterretaliation degenerate into continual conflict, but that is only one way a cyberwar can start. A further assumption is that strategic cyberwar takes place among combatants who are not fighting a real—that is, physical—war with one another, although there is some discussion in Chapter Six of the interaction between the two.

Operational cyberwar (Chapter Seven) involves the use of cyberattacks on the other side's military in the context of a physical war. The arguments here are that cyberoperations are an adjunct to kinetic operations and that success or failure at the latter determines how, when, and where conflict is resolved. Because it is nonsense to argue that a cyberattack on a military system subject to physical attack represents any kind of escalation, the assumption is that operational cyberwar has no more strategic content than the decision to introduce any other weapon in war.

Cyberdefense includes everything required to keep attackers from succeeding and benefiting from their efforts. Chapter Eight focuses on the defense of military and similar networks, which are like but not identical to civilian networks. As such, we note potential practices with especial relevance for an organization with serious, well-financed enemies; these practices include air-gapping, accountability, denial, and deception.

Chapter Nine is a brief summary.

The appendixes discuss questions raised in the monograph. Appendix A deals with the (overwrought) issue of what constitutes an act of war. Appendix B posits a model for assessing the costs and benefits of an explicit rather than an implicit cyberdeterrence policy. Appendix C looks at arms control issues.

Finally, while we hope this work has international applicability, it was written from the U.S. perspective, in particular, the perspective of a country that has invested so much in processes that depend on cyberspace. For this reason, it must wrestle with the decision of whether it should use the threat of punishment to deter the use of cyberattacks by others. The text is also written about today's capabilities and from today's perspective, one that has yet to afford enough examples of attack and counterattack to eliminate the uncertainty of what a confrontation in cyberspace would mean.

CHAPTER TWO

A Conceptual Framework

> History teaches us that a purely defensive posture poses significant risks. . . . When we apply the principle of warfare to the cyber domain, as we do to sea, air, and land, we realize the defense of the nation is better served by capabilities enabling us to take the fight to our adversaries, when necessary, to deter actions detrimental to our interests.
>
> *General James Cartwright, Vice Chairman,*
> *U.S. Joint Chiefs of Staff, 2007*[1]

Cyberspace is a thing of contrasts: It is a space and is thus similar to such other media of contention as the land and sea. It is also a space unlike any other, making it dissimilar. Cyberspace has to be appreciated on its own merits; it is a man-made construct.[2] Only after coming to such an appreciation is it possible to pick through what we believe we know about deterrence, physical warfare, and warfare in other media to figure out which elements apply in cyberspace and to what extent.

---

[1] James E. Cartwright, Statement on the United States Strategic Command Before the House Armed Services Committee, March 21, 2007. At the time of this statement, General Cartwright commanded the U.S. Strategic Command.

[2] For a broader discussion of cyberspace as such and its relationship to warfare, see Martin C. Libicki, *Conquest in Cyberspace*, Cambridge, UK: Cambridge University Press, 2007, especially the first three chapters, pp. 1–72.

The Mechanisms of Cyberspace

Cyberspace is a virtual medium, one far less tangible than ground, water, air, or even space and the RF spectrum. One way to understand cyberspace in general, and cyberattacks in particular, is to view it as consisting of three layers: the physical layer, a syntactic layer sitting above the physical, and a semantic layer sitting on top.

All information systems rest on a physical layer consisting of boxes and (sometimes) wires.[3] Remove the physical layer, and the system disappears as well. It is certainly possible to attack an information system through kinetic means, but physical attacks as such need no further elaboration here. Suffice it only to add that a computer cannot be deceived by destroying its components (although it can be through sly substitution of one component for another).

The syntactic level contains the instructions that designers and users give the machine and the protocols through which machines interact with one another—device recognition, packet framing, addressing, routing, document formatting, database manipulation, etc. Some communication infrastructures have a thicker syntactic layer than others, but every system more complex than two cans and a string has to have some. This is the level at which hacking tends to take place as human outsiders seek to assert their own authority over that of designers and users.

The topmost layer, the semantic layer, contains the information that the machine contains—the reason computers exist in the first place. Some of the information, such as address lookup tables or printer control codes, is meant for system manipulation; it is semantic in form but syntactic in purpose. Other information, such as cutting instructions or process-control information, is meant for computer-controlled machinery. The rest of a system's information is meaningful only to people because it is encoded in natural language. The distinction between information and instructions can be imprecise. Indeed,

---

[3] The link between cyberspace and electronics, although universal today, may not be universal tomorrow. Computers could be based on other principles. In the late 1960s there was some interest in using hydraulics (fluidics) as a basis for computation. The mid-1990s saw a flurry of interest in DNA computing.

many hacking tricks insert instructions in the guise of content; examples include attachments that contain viruses, overly long addresses that create buffer overflows sending the extra bits into the processing stream, and Web pages with embedded code. It is possible to attack computers solely at the semantic level by feeding them false information, like lighting a match under a thermostat to chill a room or creating a fake news source. For the most part, though, only machines whose instructions have been tampered with at the syntactic level (e.g., the user's machine has been directed to the wrong Web site or the Web site has been hacked into) will accept false information.

External Threats

Cyberattacks can be launched from outside the network, using hackers, or from the inside, using agents and rogue components. External hacking is the exemplary path for our discussion and, by far, the most common path that a state would take, particularly if going after civilian targets. Militaries and intelligence agencies, however, cannot completely ignore insider attacks, the subject of the subsequent section.

At the syntactic level, again, where hacking tends to take place, cyberspace is hedged with authorities. A person who owns a computer can normally do whatever he or she wants with it.[4] For the most part, the user should expect to retain full control over the computer, even when it is exposed to others via networking. Computers in an enterprise setting tend to come under more control by system administrators (sysadmins), and parts of such systems are closed to mere users.

To hack a computer is to violate these authorities. A hacker may send a user a rogue email or lure a user to a rogue site from which bad code is downloaded.[5] Some types of code steal information on such

---

[4] Manipulating certain forms of intellectual property contained wholly within your own personal computer is illegal under the Digital Millennium Copyright Act (especially if the intention is to break its copyright and thereby extend its useful life or permit it to be given to others).

[5] For instance, an email may purport to be from the Internal Revenue Service (see Internal Revenue Service, "Suspicious e-Mails and Identity Theft," press release, June 13, 2008).

machines.[6] Other types permit the hacker to issue subsequent commands to machine, thereby "owning" it (at least for such purposes).[7]

Hackers can also enter enterprise systems by linking to them and successfully masquerading as legitimate users with the rights and privileges of any other user. In some cases, hackers go further, fooling the system into thinking they have the privileges of sysadmins. As a sysadmin, a hacker can arbitrarily change nearly everything about a system, not least of which are the privileges other users enjoy.[8] Once hackers have wormed their way into a system and appropriated enough privileges, they can perpetrate many additional forms of mischief.[9]

The most common aim of hacking is to steal data. When states steal data from other states, it is called *computer network exploitation*

---

[6] Brian Krebs, "Virus Designed to Steal Windows Users' Data," *Washington Post*, June 25, 2004, p. A1.

[7] S. Yegulalp, "Review: Six Rootkit Detectors Protect Your System," *InformationWeek*, 2007. A hacker is said to own a machine if it can get it to do what he or she wants. This is an unfortunate use of the term *own*, which normally implies exclusivity: If I own something, you do not (co-ownership is a different matter). Rarely does so-called hacker-ownership prevent users from using their own machines. They may not even notice that their machines are doing something they did not authorize. Alas, computers are always doing something users are unaware of and would not necessarily approve of if they did know.

[8] There is a surfeit of published material about computer hacking. The more-popular items include Jon Erickson, *Hacking: the Art of Exploitation*, 2nd ed, San Francisco: No Starch Press, 2008, and Stuart McClure, Joel Scambray, and George Kurtz, *Hacking Exposed: Network Security Secrets and Solutions*, 5th ed., McGraw-Hill Osborne Media, 2005. See also the History of Computing Project, "Books on Hacking, Hackers and Hacker Ethics: An Annotated Bibliography," Web page, May 24, 2006, and the Virginia Tech Department of Computer Science, "Hacking & Security Bibliography," Web page, Blacksburg, Va., May 2, 2002.

[9] Two of the other forms of mischief, not discussed in detail here, are theft of service and unauthorized advertisement. *Theft of service* occurs when hackers run their programs on another computer's processors, ride on another network's capacity, or store materials within another computer (pornography or jihadist propaganda are two noted types of data). With everything about systems become cheaper by the year, few people worry about theft of service. *Unauthorized advertisement* (not to be confused with spyware, which leads to data theft) can be annoying but is otherwise harmless.

(CNE).[10] Corporations might also steal data (intellectual property) from other corporations. Individuals also steal data, from or, more often, about other individuals, often for the purpose of identity theft. Each may steal from the other. Because stealing data does not prevent users from enjoying free use of their own systems (in economic-speak, information is "nonrivalrous"), there may be few signs that they are being tapped into.[11] Detection is possible, if a user notices an unexpected exfiltration of data packets (although to the average user, all sorts of unexpected but perfectly legitimate exfiltrations take place); notices anomalous activities or activity patterns; notices rogue code resident on a system; or observes the consequences of a specific intrusion.[12]

Unauthorized access, however, can lead to more dastardly possibilities: disruption and corruption. Disruption takes place when systems are tricked into performing operations that make them shut down, work at a fraction of their capacity, commit obvious errors, or interfere with the operation of other systems. It is very rare that hacker attacks on code can break physical objects, but at least one laboratory demonstration of bad code caused a turbine to self-destruct.[13] Corruption

---

[10] According to a Congressional Research Service report,

> CNE is an area of Information Operations that is not yet clearly defined within DOD. Before a crisis develops, DOD seeks to prepare the IO [information operations] battlespace through intelligence, surveillance, and reconnaissance, and through extensive planning activities. This involves espionage, which in the case of IO, is usually performed through network tools that penetrate adversary systems to return information about system vulnerabilities, or that make unauthorized copies of important files. Tools used for CNE are similar to those used for CNA, but configured for intelligence collection rather than system disruption. (Clay Wilson, "Information Operations and Cyberwar: Capabilities and Related Policy Issues," Washington, D.C.: Congressional Research Service, September 14, 2006, p. 5).

[11] According to a Congressional Research Service report,

> Sophisticated attackers desire quiet, unimpeded access to the computer systems and data they take over. They must stay hidden to maintain control and gather more intelligence, or refine preparations to maximize damage. (Clay Wilson, "Computer Attack and Cyberterrorism: Vulnerabilities and Policy Issues for Congress," Washington, D.C.: Congressional Research Service, April 1, 2005, p. 37).

[12] CERT Coordination Center and AusCERT, *Windows Intruder Detection Checklist*, Pittsburgh, Pa.: Carnegie Mellon University, 2006.

[13] The Department of Energy's Idaho Laboratory undertook this event to demonstrate a vulnerability in certain supervisory control and data acquisition systems. Video was leaked

takes place when data and algorithms are changed in unauthorized ways, usually to the detriment of their correct functioning. Despite the lack of hard-and-fast distinctions between disruption and corruption, a good rule of thumb is that the effects of disruption are drastic, immediate, and obvious, while the effects of corruption are subtle and may linger or recur. It is relatively easy to tell that a system is not working. It is harder to tell that it functions but generates wrong information or makes bad decisions.[14]

Hackers intent on causing later mischief often facilitate their efforts by dropping rogue computer code into systems for later use.[15] What can be termed implants often lie dormant, only to be activated either by events on the target machine (e.g., the appearance of a new information of the sort the hacker might be interested in) or by commands from the hacker. In some cases implants operate autonomously, searching for computers on the network that lack such implants and making sure they do not lack for long.

Regardless of what the hacker intends to do to steal information— disrupt systems or corrupt them—the first, and often the most difficult step, is, in fact, getting inside (that is, receiving the privileges of a system's user or administrator). For this reason, the early phases of a CNE look the same as the early phases of a computer-network attack. As a corollary, those with the best capability to get inside another system tend to be best qualified to carry out a computer-network attack.

Because the syntactic layer rests on the physical layer, one can confidently assert that there is no forced entry in cyberspace.[16] If someone

---

to CNN (Jeanne Meserve, "Sources: Staged Cyber Attack Reveals Vulnerability in Power Grid," *CNN.com*, September 26, 2007).

[14] This requires some a priori way of knowing what the right information or decision is—and if that were that easy, who would need computers? Consider the possibility that the guidance-and-control system on a new missile might have been tampered with. If the missile lacks a good track record, it might be difficult to know whether it failed because the software had errors or because its software had been altered, especially if the actual missile were unrecoverable.

[15] Wilson, 2005.

[16] One can forcibly put a component or a box into a network, but this requires disguising the fact of the forcible entry. In a network with RF links, one signal can overpower another

has gotten into a system from the outside,[17] it is because that someone has persuaded the system to do what its users did not really want done and what its designers believed they had built the system to prevent. Nevertheless, in any contest between a computer's design and use-model (e.g., a user's intuition that email is information, not instructions) on the one hand and its software code, on the other, the code always wins. Whoever gets into a system gets into it through paths that the software permits.[18] The software may have flaws (most would be inadvertent, but some may be deliberate; see below) or may have been misconfigured (e.g., the permissions the administrator established differ from the permissions that the administrator thought had been established), but a system is what it is, not necessarily what it should be.[19]

---

and thereby provide an entry point for errant bytes—although if the original signal were not nulled out, the fact that a signal has been overridden should be detectable.

[17] Users are supposed to get into systems, but obtaining a password can allow the wrong someone to pose as a particular user.

[18] The term *software* is used as shorthand for *instructions*. Some instructions are hard-coded into the firmware or hardware.

[19] There is one exception to the rule that computer attacks arise from host-system vulnerabilities. In a *flooding attack*, a hacker generates such a high volume of packets destined for a particular network location that legitimate information does not get to the affected network, cutting it off from the rest of the world. These days it is very difficult for one computer to flood another.

Floods today thus tend to arise from DDOS attacks, generally caused by thousands or millions of unwitting computers (bots) that have come under the control of hackers. There is currently a thriving business in creating large numbers of bots (by subverting third-party computers), organizing them into botnets, and renting their services to others, such as spammers. The first well-known DDOS attack occurred in February 2000, when several e-commerce sites were taken down for periods ranging up to several hours. The attack on Estonia was of this type. By one estimate, up to one in ten packets over the Internet is part of some bot attack (Robert Lemos, "A Year Later, DDOS Attacks Still a Major Web Threat," *CNET News*, February 7, 2001).

DDOS attacks are difficult to defeat precisely because they can target otherwise well-protected networks. The programs that convert third-party computers into bots only need to work against the least well-protected Internet-linked computers to take them over. Spamming aside, most bot attacks affect people in shady businesses, so there is little political will to shut down access to machines whose innocent owners have unknowingly let them be converted into bots. This could change if DDOS attacks become a far more serious problem to the Internet than they have been to date. Conceivably, Internet service providers could be ordered by law to cut off service to users whose machines demonstrate an unusual pattern of

Such a divergence, when it has security implications, is a vulnerability. Whatever the methods, manual or automated, hackers use, an attempt to take advantage of a vulnerability to gain access to a system or to get it to accept rogue instructions is called an *exploit*.

A system's integrity dictates how badly a system can be hurt by attacks in cyberspace. One might even argue that a system's integrity is a more important determinant of success than the quality of the adversary's exploits—after all, no vulnerabilities, no exploits; no exploits, no cyberattacks.[20]

Thus, in theory, all computer mischief is ultimately the fault of the system's owner—if not because of misuse or misconfiguration, then because of using a system with security bugs in the first place. In practice, all computer systems are susceptible to errors. The divergence between design and code is a consequence of the complexity of software systems and the potential for human error. The more complex the system—and they do get continually more complex—the more places there are in which errors can hide. Every information system has vulnerabilities—some more serious than others. The software suppliers themselves find a large share of these vulnerabilities and issue periodic patches, which users are then supposed to install—which some do more expeditiously and correctly than others.[21] Hackers find some vulnerabilities and then spring corresponding exploits on unsuspecting users who have otherwise done everything correctly. Literally thousands of exploits are sitting around. Many of the more-devious ones

---

outgoing packets. Internet service providers could force users to demonstrate the presence of antivirus software before allowing them on the networks. To work, however, such solutions would have to be applied worldwide—an unlikely prospect.

Thus, for the time being, DDOS attacks are likely to remain a threat. Fortunately, DDOS attacks cannot corrupt or, these days, crash systems, and they do not affect traffic internal to server-restricted spaces. (See also Clay Wilson, "Botnets, Cybercrime, and Cyberterrorism: Vulnerabilities and Policy Issues for Congress," Washington, D.C.: Congressional Research Service, January 29, 2008, p. 25.)

[20] Rogue users or, worse, sysadmins, present a vulnerability of a different sort; see the section on internal threats that follows.

[21] Unfortunately, hackers carefully observe patch releases and often reverse engineer them, determining the vulnerabilities the patches were supposed to fix, developing appropriate exploits, and using the new exploits against those who have not yet installed the patch.

require physical access to the target system. Most of the ones that reach the news do not work on well-patched systems.

In a sense, cyberattacks rely on deception—persuading systems to do what their designers do not want them to do. Fortunately, deception can be its own undoing. An exploit, if discovered, shows sysadmins that something is not right. With good logs, sysadmins may be able to determine where something unusual took place in the interaction between the hacker and the system.[22] Changes in files (data or instructions), or the presence of unexpected files can also be telling. The process is hardly perfect; it is possible to determine a specific vulnerability and miss the broader design flaw of which the specific vulnerability is just an instance. Furthermore, individual system administrators almost never have direct visibility into packaged software and cannot fix vulnerabilities of which the software vendor is itself unaware. Nevertheless, any one sysadmin can take advantage of an international community with a common interest in minimizing outstanding vulnerabilities.

In contemplating cyberspace, it may help to differentiate the peripheries of the system. The peripheries may be said to contain user equipment; that is, equipment whose function and parameters are established by users.[23] Peripheries, if not air gapped or protected via consistent encryption, tend to be repeatedly vulnerable largely because users are rarely trained in or focused on information security.[24] User systems and privileges can be taken over through password cracking, phishing, social engineering, downloads from bad Web sites, use of bad media (e.g., corrupted zip drives), etc. Sadly, the security of the periphery as a whole is often no better than the security of the most feckless user. The core, by contrast, is what sysadmins control—

---

[22] CERT Coordination Center and AusCERT, 2006.

[23] A personal computer is within the periphery, while an old-fashioned dumb terminal (e.g., an IBM 3270 terminal) would have been part of the core, because users cannot do anything on the latter except scripted data entry. In practice, the distinction is somewhat fuzzy; users could command DEC's VT100 terminal in ways that weakened the systems it accessed.

[24] An *air gap* is the lack of an electronic connection between the system and the rest of the world. A true air gap also requires that the system not have RF links (or at least not ones powerful enough to be picked up beyond the protected perimeter), and that no media (say, thumb drives) cross the perimeter (such as by being removed from a computer).

monitors, routers, management devices, machinery (such as weapons), and databases. Sysadmins are (or should be) trained and sensitive to security issues; they also set the terms by which users (and their systems) interact with the core. Although it is good personnel practice to sensitize users to security issues, it is good engineering practice to assume that users will not always be sensitive. While it is possible to protect the core from insecure users, it is less clear whether networks can function when enough user systems are compromised badly enough, even though network administration is a function of sysadmins. In general, it is hard to compromise the core in the same precise way twice, but the periphery is always at risk.

Internal Threats

States have two other methods of gaining access to systems; in fact, these are the only ways to get into truly closed systems. One is to recruit insiders, who, with varying degrees of help, can introduce mischief into systems (especially if they are sysadmins themselves). The other is to toy with the supply chain so that target systems contain components that appear benign but contain code that responds to a state's directions or at least priorities.[25] Unlike computer hacking, many of whose techniques are published on the Web and in print, the insider and component methods are essentially the province of state intelligence agencies and therefore highly protected. It is unclear how well they have worked.

Insiders
Unlike operating a system connected to the rest of the world, which is known to contain hackers, operating one from the inside to create mischief tends to violate explicit trust conditions. System operators capable of doing so must go through much more comprehensive security processes to achieve any given level of security in the face of an insider threat.

---

[25] Although the components themselves are often hardware, it is almost always the software (e.g., the firmware, the hardwired microcode) that is altered.

The insider threat has always been a staple of computer security, not least in the banking industry.[26] Most well-managed systems therefore make it difficult for rogue employees to do a great deal of damage and, in some cases, limit how much material they can access (even when any individual item is available). As a general principle, a rogue employee presents risks similar to those of a feckless user in the periphery of an open system. Rogue sysadmins are a much deeper headache but, again, one that banks deal with constantly. Insider recruitment can produce unexpectedly sharp pain but does not lend itself to nationwide effects very easily. Such an attack cannot be duplicated at will, as an exploit can (unless used as a launch point for a hacker or malware attack that cannot otherwise work as well). Revealing one turncoat can lead to investigations that could unravel entire recruitment networks.

Supply Chain
Notable cases of successfully compromised components include (1) the British donation of Enigma machines to other nations, which likely did not realize that the British were able to break and thereby read messages from such machines,[27] and (2) the installation in the Soviet natural gas network of (black-market) system controllers altered to malfunction in ways that lead to destructive pipeline explosions.[28] There are also suspicions that some cryptographic devices a Swiss company sold had a

---

[26] Marissa Reddy Randazzo, Michelle Keeney, Eileen Kowalski, Dawn Cappelli, and Andrew Moore, *Insider Threat Study: Illicit Cyber Activity in the Banking and Finance Sector*, Pittsburgh, Pa.: CERT Coordination Center, Software Engineering Institute, Carnegie Mellon University, June 2005.

[27] According to Simon Singh,

> Britain had captured thousands of Enigma machines and distributed them among its former colonies, who believed that the cipher was as secure as it had seemed to the Germans. The British did nothing to disabuse them of this belief, and routinely deciphered their secret communications in the years that followed. (Simon Singh, *The Code Book: The Evolution of Secrecy from Mary Queen of Scots to Quantum Cryptography*, New York City: Random House, 1999, p. 187.)

[28] Thomas C. Reed, *At the Abyss: An Insider's History of the Cold War*, San Francisco: Presidio Press, 2005.

National Security Agency (NSA)–sponsored back door.[29] Many in the defense community worry that China's growing presence in component manufacturing provides it plenty of opportunities for mischief—which it may not be shy about taking advantage of.

Unless and until purchasers get access to all the code in the electronics they buy, a supply-chain attack is difficult to defend against.[30] Such components can fail and perhaps bring down the rest of the system at a prespecified time or in response to some system state. Yet there are also limits to what rogue components can do if installed in a truly air-gapped system: They cannot respond to a signal unless the system can receive messages from the outside,[31] and they cannot exfiltrate information unless the system can generate messages. A component attack is a sporty move. The discovery of lead paint contamination on Chinese toys led to major tremors in China—and that was just sloppiness.[32] Consider what damage a deliberately corrupted component would have on China's reputation, much less the reputation of the guilty supplier. One discovery may create the incentive to recycle everything acquired from the now-suspect source.

In Sum

Neither agents nor corrupted components violate the basic tenets discussed earlier. Neither represents forced entry. Both are forms of deception and of the sort that the once-deceived is unlikely to fall for as easily again. In contrast, however, to the (theoretical) presumption that

---

[29] As a small company, the supplier stood to lose little other legitimate business, and cryptography is an area in which the benefits from corrupting a device are obvious. See Scott Shane and Tom Rowman, "Rigging the Game," *Baltimore Sun*, December 10, 1995, p. 1A. For the company's denial, see Scott Shane and Tom Rowman, "Congress Has Tough Time Performing Watchdog Role," *Baltimore Sun*, December 15, 1995, p. 23. See also Ludwig de Braeckeleer, "For Years US Eavesdroppers Could Read Encrypted Messages Without the Least Difficulty," *The Intelligence Daily*, December 29, 2007.

[30] Sally Adee, "The Hunt for the Kill Switch," *Spectrum*, May 2008.

[31] Or from a rogue insider. Note that the component itself may break the air-gap status of the system it is in if it has hidden transmit-receive circuitry.

[32] David Barboza, "Owner of Chinese Toy Factory Commits Suicide," *New York Times*, August 14, 2007.

a system cannot be fooled the same way twice, the high cost of validating components and employees suggests that rogues of either sort can recur—but even a wily attacker will face serious obstacles to repeating success.

Defining Cyberattack

With that as background, *cyberattack*, for the purposes of this discussion, is the deliberate disruption or corruption by one state of a system of interest to another state. The former state will be referred to as the *attacker*; the latter state will be referred to as the *target*. In some contexts, the target may also become a retaliator. The affected system will be referred to as the *target system*. Appendix A addresses whether a cyberattack constitutes an act of war.

Note a key ramification of this definition: CNE (spying) is not an attack (as disruption and corruption are). Note as well two assumptions: The attacker is a state and the target is a system of interest to another state. Why make such distinctions?

CNE deserves to be distinguished from cyberattack. First, CNE does not deprive the user of the full use of the machine. The user suffers no consequential harm other than having secrets stolen. Second, because CNE is so difficult to detect, a deterrence policy could only be activated by exception. Harsh punishments for crimes that are rarely detected tend to lose credibility as law enforcement mechanisms, and this is even more true if such methods are used to try to govern the activities of other states.[33] Third, the law of war rarely recognizes espio-

---

[33] Gary S. Becker provides the classic formulation of the position that the odds of punishment rather than its severity reduce criminal activity in "Crime and Punishment: An Economic Approach," *The Journal of Political Economy*, Vol. 76, No. 2, 1968, p. 9:

> [A] common generalization by persons with judicial experience is that a change in the probability has a greater effect on the number of offenses than a change in the punishment, although, as far as I can tell, none of the prominent theories shed any light on this relation. [footnote:] For example, Lord Shawcross (1965) [Lord Shawcross, "Crime *Does Pay Because We Do Not Back Up the Police*," *New York Times Magazine*, June 13, 1965] said, "Some judges preoccupy themselves with methods of punishment. This is their job. But in preventing crime it is of less significance than they like to think. Certainty of detection is far more important than severity of punishment."

nage as a *casus belli*, and a good case for changing this has yet to be made, even though the means of espionage have changed.[34] Fourth, everyone does it.[35] Those who try to establish deterrence policies to prevent others from doing what they do themselves perforce reveal themselves to be fools or hypocrites—unless they are so powerful that they can get away with it. It is doubtful whether even the United States qualifies as being that powerful. A deterrence posture against CNE would be viewed as hypocritical and probably not credible—indeed, as incredible.

That stated, CNE is often called an "attack."[36] Furthermore, it is entirely plausible that the victim of a large-scale attempt at espionage (e.g., a wave of pings) might believe that it is seeing the preparations for a full-scale attack as defined here.[37] An implant designed to purloin information may be indistinguishable from an implant designed to disrupt systems or corrupt information. A jumpy defender may react to

---

[34] In the beginning of the space age, President Dwight Eisenhower declared that a satellite transiting over another state was not an act of trespass. Although this was necessitated by the laws of orbital mechanics (most low earth orbits could not avoid going over the Soviet Union because of its size), the political motivation was to establish legal precedent for spy satellites. It helped, politically, that the Soviets initially could do very little about our space activities and, a few years later, realized the advantages of being able to do the same thing themselves. See Walter McDougall, *The Heavens and the Earth: A Political History of the Space Age*, New York: Basic Books, Inc., 1985, pp. 121–124.

[35] For instance, the Germans; see Gadi Evron, "German Intelligence Caught Red-Handed in Computer Spying, Analysis," blog post, *Security Dark Reading*, Web site, March 11, 2009. See also Vernon Loeb, "Test of Strength," *Washington Post Magazine*, July 29, 2001, p. W08. Also, according to the Congressional Research Service,

> U.S. intelligence officials, speaking on background, explained that they have routinely penetrated potential enemies' computer networks. These officials claim that thousands of attacks have taken place and sensitive information was stolen. (Wilson, 2008, p. 12)

[36] Consider, for instance, the material in Grow, Epstein, and Tschang, 2008.

[37] Melissa Hathaway, senior advisor to the Director of National Intelligence, observed in late 2008 that

> we are finding a persistent presence on these networks and we cannot say with assurance that a network that has been penetrated has not been infected with hidden software that could be triggered in a crisis to disrupt or destroy data or communications. (Melissa E. Hathaway, "Cyber Security: An Economic and National Security Crisis," *The Intelligencer: Journal of U.S. Intelligence Studies*, Vol. 16, No. 2, Fall 2008, p. 31)

an attempt to steal information as if beset by an attempt to do damage, and that reaction may be a counterattack. Chapters Three and Four detail some real risks of mischaracterization and misattribution.

True, a great deal of state-sponsored CNE is going on. The PLA stands accused of having broken into thousands of civilian and unclassified military systems, in the United States and elsewhere (e.g., Germany), to steal large quantities of information.[38] The Chinese are also said to have dropped implants into such systems in ways that make it difficult to clean up individual machines without allowing them to become reinfected.[39] Germany's chancellor, Andrea Merkel, felt confident enough in this attribution to complain to China's premier in person.[40] China has steadfastly denied all responsibility.[41]

Restricting our discussion to states is a requirement for talking about deterrence in kind. Retaliation in kind is possible only if it is possible to place the information systems of value to the attacker at

---

[38] In August 2007, on the eve of German Chancellor Angela Merkel's meeting with Chinese Premier Wen Jiabao in China, *Spiegel* magazine reported that Germany's domestic intelligence service, the Office for the Protection of the Constitution, had discovered a significant cyberattack targeting computers in the German Chancellery and in the foreign, economic and research ministries in May 2007 ("Merkel's China Visit Marred by Hacking Allegations," *Spiegel Online International*, August 27, 2007). In this instance, the information was siphoned off the German government's machines using Trojan horse programs that sent German government data via the Internet to what is believed to be a PLA-supported locus of attack in Lanzhou, Canton province, and to Beijing. While the German government does not know exactly how much information was stolen, some estimates are in the terabytes, and German security officials were able to thwart a 160-gigabyte data transfer. German security officials also said they estimate 40 percent of all German companies have been victims of nation-state–sponsored industrial espionage, with the majority of the activities originating in Russia and China (Christopher Burgess, "Nation States' Espionage and Counterespionage: An Overview of the 2007 Global Economic Espionage Landscape," *CSO Online,* April 21, 2008).

[39] There is no word on whether these implants could have launched or facilitated a destructive cyberattack or were instead simply meant to facilitate further espionage.

[40] Rogers Boyes, "China Accused of Hacking into Heart of Merkel Administration," *Times Online* (London), August 27, 2007; "Merkel's China Visit Marred by Hacking Allegations," 2007.

[41] See Nathan Thornburgh, "Inside the Chinese Hack Attack," *Time,* August 25, 2005a, for a 2004 report; Grow, Epstein, and Tschang, 2008, for a 2006 report; and John Blau, "German Gov't PCs Hacked, China Offers to Investigate," *PC World,* August 27, 2007.

risk. True, any state that has a telephone system, especially a cell phone system, has something at risk, and any state that has purchased a turnkey facility from any corporation has probably purchased a complex information system to run it. Yet nations vary greatly in the degree of their dependence on information systems and hence in the degree to which disruption of any given state's systems may harm it. It would be conceivable to extend the domain from states to well-wired corporations (or similar enterprises) with systems at risk, but corporations rarely number among the hackers, and their assets are almost always subject to the laws of governments. Therefore, nothing of analytic substance would be gained by including corporations. Although well-funded groups (as Aum Shinrikyo was 15 years ago) could engineer cyberattacks as well as most states, they, too, rarely have information systems at risk. Violent islamic fundamentalists, for instance, tend to skim off established networks or rely on the help of their friends. Unless such groups enjoy quasi-sovereign status somewhere, criminal prosecution (or extrajudicial equivalent) will have to do.

Clearly, an attacker without an information system of its own is unlikely to be deterred by retaliation in kind (cyberattacks that, for instance, wipe out the attacker's bank account may be dissuasive, but the ramifications for the bank, a presumably innocent third party, are more serious).[42] This does not, however, mean that cyberattacks by individuals and nonstate actors cannot be deterred by other means, and these means may require cyberforensics (to establish culpability) or other intelligence gathered through cyberspace. We choose, however, not to define the use of cyberspace to support other forms of deterrence as deterrence in kind.

The last criterion, that the target system has to be of interest to a state, is fairly self-evident, but it raises a question: Can a state legitimately threaten retaliation for attacking an information system outside its borders? The first short answer is a question: Why not? Imagine an attack on the information system of a credit-card company that disables

---

[42] Nonstate actors may have Web sites that can be turned off by hacking, or more lawfully, by persuading its host to end their service. Al Qaeda's Web sites are constantly subjected to such treatment; it makes their propaganda activities more difficult but hardly impossible.

all such transactions for an extended period; should it matter than the critical computer was located in Canada rather than the United States? Indeed, who, other than its owners actually knows where the relevant servers sit? The second short answer, however, is not to leap so fast. Set aside the question of whether a state should retaliate for any attack on a private system. The practical difficulty remains of getting the affected owner of a hacked system to reveal enough technical information to figure out what happened and who did it. A state that would retaliate would likely find it more difficult to get deep access into attacked systems if the systems lie outside its borders. Such difficulties are not insurmountable, but the following discussion makes no less sense if we ignore such cases.

Defining Cyberdeterrence

We have chosen to define cyberdeterrence as deterrence in kind to test the proposition that the United States, as General Cartwright offered, needs to develop a capability in cyberspace to do unto others what others may want to do unto us.[43] The need for such a capability has to assume that there will be times when something more violent and more attention-getting is off the table. Otherwise, why bother with something weaker? Not surprisingly, communities contemplating or developing a cyberattack capabilities have a greater interest in cyberdeterrence than communities armed with a conventional response capability. For our purposes, assuming retaliation in kind raises almost all the issues that more-violent forms of retaliation do, as well as many issues that the latter do not.

---

[43] This is not just an American sentiment. From Indrani Bagchi, "China Mounts Cyber Attacks on Indian Sites," *The Times of India*, May 5, 2008:

> A quiet effort is under way to set up defense mechanisms, but cyberwarfare is yet to become a big component of India's security doctrine. Dedicated teams of officials—all underpaid, of course—are involved in a daily deflection of attacks. But the real gap is that a retaliatory offensive system is yet to be created. And it's not difficult, said sources. Chinese networks are very porous—and India is an acknowledged information technology giant.

Those who believe that retaliation should be controlled to avoid escalation and that the use of physical forces represents a retaliation in kind over the use of cyberattacks may find that retaliation in kind is more attractive than more-violent alternatives because the former raises fewer issues of proportionality. Conversely, retaliation in kind may legitimize a form of warfare that it would not be in the interest of the United States to legitimize when it has more than adequate conventional strength for every occasion. By contrast, the United States is at least as vulnerable to such attacks as any other state and is far more dependent on information systems than current competitors (e.g., China) and lesser threats (e.g., Iran).

Likewise, this definition of cyberdeterrence does not include such lesser measures as prosecuting hackers themselves or taking diplomatic or economic measures,[44] although Chapter Five does touch on lesser measures. This is not to say that the threat of such measures cannot dampen the other side's aggressiveness in cyberspace—they may, in fact, be the better course of wisdom. However, if something less aggressive sufficed, why bother with cyberdeterrence? Again, expanding the definition is unnecessary for tackling the many issues discussed here. Similarly, for purposes of discussion, this definition does not entail cyberresponses to other types of attacks (although Chapter Four does touch on this). Figure 2.1 depicts four types of response, listed in rough order of level of belligerence (although not necessarily the magnitude of the consequences).

The aim of deterrence is to create disincentives for starting or carrying out further hostile action. The target threatens to punish bad behavior but implicitly promises to withhold punishment if there are no bad acts or at least none that meet some threshold. At a minimum, it requires the ability to distinguish good behavior from bad. False positives and false negatives are both bad for such a policy, but the former is worse. Undeserved punishment lacks legitimacy: If the presumed attacker is innocent, the retaliator may have made a new enemy. Even

---

[44] These are only approximate measures of belligerence. A general embargo on a state might be appear to be a lot more belligerent than a cyberresponse the state might not care much about.

Figure 2.1
Responses by Rough Order of the Level of Belligerence

More belligerent

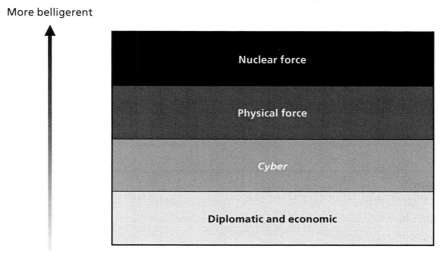

Less belligerent
RAND *MG877-2.1*

if the presumed attacker deserved punishment for other bad behavior, its incentive to behave well may be attenuated: If punishment follows both innocence and crime, why avoid crime? Failure to punish the guilty weakens deterrence but does not necessarily eliminate it; much depends on the likelihood and consequences of getting caught. People are inhibited from committing crimes even though the odds of getting caught are less than 100 percent—if they figure that the consequences of getting caught greatly exceed the criminal gains. To some extent, if the odds of getting caught are low, the potential retaliator can keep the expected cost of being caught constant by keeping punishment high. This is true, however, only if punishment is within reason: Rare severe punishments tend to be perceived as disproportionate and hence less legitimate. If one is operating within a peer-level system, outsized response may be interpreted as aggressive acts of escalation.

Deterrence also requires the adversary to be able to distinguish being punished from not being punished. In most realms this is not a problem—not so in cyberspace. As discussed further below, not only might both the retaliator and the attacker be unable to predict the effect of retaliation, neither may be entirely certain of what effect retaliation did have. If the potential retaliator doubts whether its planned retaliation will have the desired effect, it may be better off pretending that no attack occurred (quite possible in some cases) than making a big deal of the attack, revving up the retaliation machine, and having little or nothing to show for it.

Deterrence in general comes in many forms. Some forms are singular, while others have to be repeated. Some are asymmetric and some are symmetric (among peers).

Nuclear deterrence is singular and symmetric. It is singular in that the point is to make the prospect so frightening that no one dares invoke it. If nuclear retaliation ensues, by the time retaliation and counterretaliation has run its course, the (literal) landscape and hence the strategic circumstances underlying the deterrence are likely to have become quite different. One or both parties may have been eliminated, lost their freedom of action, or been rendered powerless. The nature of deterrence the second time will also be different. The same largely holds true for heavy conventional deterrence: If retaliation is invoked, it too is likely to run its course and lead to a major war, which one or another regime also may not survive.[45]

Criminal deterrence is repeatable and asymmetric. It has to be repeatable because many first-time offenders become second-time offenders. The prospect of counterretaliation from criminals, meanwhile, is not a serious problem in the United States and developed countries in general. Police and other officials of the justice system are rarely at personal risk, thanks in large part to the legitimacy they are accorded and their latent ability to mass force and the force of law against criminals. Communities in which this is not the case (e.g.,

---

[45] See, for instance, John J. Mearsheimer, *Conventional Deterrence*, Ithaca: Cornell University Press, 1983. George Quester argued that the British were deterred—more precisely, inhibited—from intervening against Hitler in 1938 because they feared the Luftwaffe (whose size they grossly overstated). Once engaged in war, they learned that defenses against air attack were, in fact, possible; as a result, they lost fewer lives during the Blitz than many Axis cities did in overnight raids the Allies subsequently conducted (George Quester, *Deterrence Before Hiroshima*, New Brunswick, N.J.: Transaction Books, 1986).

drug-infested states in the rest of the Americas, formerly insurgent-dominated precincts in Iraq) are clearly troubled. There, criminal justice has been rendered ineffective, and the rule of the jungle prevails.

Cyber deterrence has to be repeatable because no feasible act of cyberretaliation is likely to eliminate the offending state,[46] lead to the government's overthrow,[47] or even disarm the state. Thus, a state could attack, suffer retaliation, and live to attack another day. But cyber deterrence is also symmetric because it takes place among peers. The target state (the potential retaliator) does not, a priori, occupy a higher moral ground than the attacker. There is also no reason to believe that the target can win any confrontation with the attacker if things go too far. Thus, the retaliator always has to worry about counterretaliation (as it does in nuclear conflict) and cannot help but shape its deterrence policy with that in mind.

Cyberdeterrence is not unique in being repeatable and symmetric. Such deterrence typically characterizes interactions among quarreling states (or quarreling tribes for that matter), each on guard against depredations from the other side and each willing to defend itself against small slights. Deterrence in such situations does not necessarily keep the peace; in an anarchic system, violence is endemic. Fights, in retrospect, often look like they have no larger issues than themselves.

Given its conventional military power, the United States enjoys the kind of superiority that permits it to be the global cop, on the lookout for bad behavior without worrying terribly much about how others may react. This is not the situation in cyberspace. The United States may have superior offensive capability—having invested large sums in

---

[46] Cyberattacks can eliminate smaller entities in the sense of putting them out of business. Blue Security, a firm whose business was to block spam, ran afoul of one such spammer, who went under the *nom du hack* of PharmaMaster. The latter engineered a DDOS attack that flooded Blue Security's intakes long enough for Blue Security to decide that the fight and collateral damage were too much. The company agreed to shut down its antispam service. In that sense, PharmaMaster won that cyberwar.

[47] Conceivably, the government could turn over peacefully in the wake of the poor choices that lead to a cyberwar, although it is putatively more likely that a conflict would initially strengthen the government's political standing. Goading Israel into attacking Lebanon did not seem to harm Hezbollah's status in Lebanon any.

such capabilities—and the state's code writing talents are second to none—software is one sector where the United States is still a clear net exporter. But the United States is also quite vulnerable. Its society and, in particular, its military (with its attraction to network-centric warfare) depend heavily on information systems. The systems that are run both privately and publicly tend to be more accessible, at least compared to those of less well developed or less democratic nations. There is also very little uniformity in security policies across the many U.S. institutions. This may make society more robust overall, but it also presents more opportunities for mischief and introduces security seams where different organizations connect. More to the point, it is precisely because others suffer inferiority in conventional conflict that they feel driven to emphasize cyberattacks as a way to even the score. Thus, the United States, for all its advantages, might suffer more than adversaries would if retaliation begets counterretaliation.

Cyberdeterrence, for its own part, is a policy that the United States has a choice about adopting. Bear in mind what the goal is: reducing the risk of cyberattacks to an acceptable level at an acceptable cost. If cyberdefense can suffice for that, why run the additional risk of threatening a confrontation to protect systems? Unfortunately, information security can be quite expensive. The expenditures of U.S. organizations on information security easily measure in the tens of billions of dollars a year—yet security breaches occur daily.[48] This is why the federal government sought $7.3 billion in fiscal year 2009 to protect government computers over and above the billions already spent.[49] Offense-defense curves at levels that characterize today's cyberspace favor the offense. That is, another dollar's worth of offense requires far more than another dollar's worth of defense to restore prior levels of security.[50] This is illustrated in Figure 2.2 (at this point, pay attention only to

---

[48] Dawn S. Onley, "Army Urged to Step Up IT Security Focus," *Government Computer News*, Vol. 1, No. 1, September 2, 2004.

[49] Wyatt Kash, "Spending for IT Security Gains Ground in 09 Budget," *Government Computer News*, February 7, 2008.

[50] The sharp-eyed reader may note that this disparity—each marginal dollar spent on offense is offset by several dollars worth of defense—appears to contradict the economic tenet that

Figure 2.2
Does the Cost-Effectiveness of a Cyberattack Decline at High Intensity Levels?

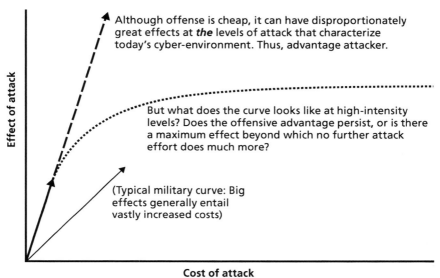

Although offense is cheap, it can have disproportionately great effects at *the* levels of attack that characterize today's cyber-environment. Thus, advantage attacker.

Effect of attack

But what does the curve looks like at high-intensity levels? Does the offensive advantage persist, or is there a maximum effect beyond which no further attack effort does much more?

(Typical military curve: Big effects generally entail vastly increased costs)

Cost of attack

RAND *MG877-2.2*

the lower left-hand side of the solid line).[51] Part of the reason offense is rather inexpensive at today's levels is that individual hackers are rarely at personal risk. This gives them incentive to push governments to keep going—or privatize their efforts if the individual's own government is squeamish about the cost (but not hostile enough to arrest hackers).

The attraction of cyberdeterrence is that, if it works, it can reduce the cost of defending systems. Instead of having to put money into

---

an organization invest in something only up to the point that a dollar spent equals a dollar's worth of value. The reason an attacker does not care to spend as much on offense as on defense, such that the two are equal, may be that it does not value its target's pain as highly as it values its own cost of causing that pain. An attacker may figure that, if expending one dollar of additional effort does not produce at least ten dollars worth of damage to the target, the effort is not worthwhile.

[51] Think of the "Cost of attack" in Figure 2.2 as the least-cost combination of the damage from the attack and the cost of security required to reduce the damage from what it could have been to the actual level.

making systems more secure, the defender inhibits the attacker's efforts by threatening retaliation against successful attacks (or, if the defender is sufficiently confident in its forensics, against robust attempts to attack). If the attacker can be persuaded to reduce its efforts in the face of punishment, the defender can save some of what it would have spent on defense and still achieve the same level of security.

The amount of money that even a successful cyberdeterrence policy can actually save should not be overstated. The bulk of computer security would still have to be in place to deal with rogue individuals and nonstate threats. There are no vulnerabilities that a state could discover that an individual cannot discover and exploit. Thus, there are no defensive measures against hacking that can be dispensed with, even if all states were deterred.[52] Furthermore, it takes more than credibility to realize the payoff from a cyberdeterrence policy. Even if an attacker believed getting caught would have grave consequences, it may not believe it will be caught; if caught, it can put more effort into hiding better next time.[53] Finally, those who advocate greater cyberdefense generally also advocate greater cyberdeterrence; the political debate is not an either-or proposition. The real division is between those who are more alarmed at the threat and those who are less alarmed. If the

---

[52] States *do* have an advantage when it comes to attacking systems in two other ways. States are more practiced at recruiting individuals, but they are not the only source of insider threats that must be protected against. They also have the resources to substitute altered components for the real thing. This is a form of attack hard to imagine individuals or groups pulling off. A successful cyberdeterrence policy may allow some slight relaxation about the threat to components—something few outside the intelligence community worry about in any case.

[53] The argument that punishing mischief in cyberspace only drives attackers to hide themselves better smells a lot like the argument that additional defenses only drive attackers to surmount them. Thus, if punishment is pointless (because hiding easily nullifies it), so is defense (because greater sophistication easily nullifies it). That argument loses force, however, if attackers do *not* fear punishment and therefore take few pains to avoid attribution. Conversely, there are low-cost measures they can take to hide themselves, *if* punishment becomes an issue. The only parallel that applies to defenses is attacks that take place against systems whose owners do not care (which is why so many of the attempts to create bots work). If there are, in fact, well-defended systems that face sophisticated foes with an obvious motive to crack them—banks and militaries are surely in that category—the attackers have already tried (or dismissed as pointless) all the easy steps to get in. Yet such systems continue to function; ergo, the defenses must have done something useful.

United States begins to put serious resources into cyberdeterrence, it may be assumed that the resources the government invests in cyber-defense are also going up.

The other reason to doubt that a cyberdeterrence policy could markedly reduce what the United States spends today on defense is that the country has yet to be attacked—or at least seriously and success-fully attacked—by another state in cyberspace. Thus, it may already be profiting from an implicit deterrence stance that warns other states against any seriously hostile act. This is not to say that state hackers have given U.S. systems a pass. They have undoubtedly penetrated U.S. systems, if only to steal information (as far as anyone knows), some-thing unlikely to be actionable in any reasonable deterrence policy. But absent serious attacks, the possibility remains that adding a threat to retaliate in cyberspace offers no upside in terms of reducing today's attacks.

A better case for deterrence presumes potential adversaries have preparations for cyberattacks that go much farther and are much more systematic than anything we have seen to date. This would mean that the current paucity of attacks has nothing to do with the fear of retalia-tion and more to do with the inability of other states to generate a suffi-ciently interesting capability or a sufficiently pressing opportunity—so far. What, then, would it take to retain an adequate level of security in the face of potentially more intense attacks?

If the offense-defense curves continue to favor the offense, one could argue that either the potential damage from a cyberattack would be unacceptable or the resources that must be spent on defense are unaffordable. The United States therefore has no recourse but to hit back after the fact. Rather than let matters get to that pass, the argu-ment goes, the United States should make clear to potential attack-ers that they will be counterattacked in kind. This would affect their calculations today over what to invest in such a capability. If they are dissuaded today, the odds of a full-fledged conflagration tomorrow go down.

It is entirely possible, however, that the offense-defense curves (see the dotted line in Figure 2.2) would flatten out at higher levels of inten-sity. If sufficient expenditures are made and pains are taken to secure critical networks (e.g., making it impossible to alter operating parameters of electric distribution networks from the outside), not even the most clever hacker could break into such a system. Such a development is not impossible. Given the many risks involved in starting a fight over problems that can be managed simply by making more effort, the case for deterrence would be considerably weakened.

But no one knows what the curves are at intensity levels significantly higher than those seen today.

In practice, the difference between deterrence advocates and deterrence skeptics is less one of doubt about the shape of future curves and more about differences on the level of current damage.[54] Cyberhawks, as it were, maintain that the damage is much worse than we know. They argue that organizations struck by hackers are reluctant to confess that they were hacked, lest they look feckless.[55] When they do go to the FBI, everyone involved keeps quiet. Damage to sensitive information systems could be much greater than even the owners know because a great quantity of rogue code could be hiding, awaiting an activation

---

[54] CSIS Commission on Cybersecurity for the 44th Presidency, *Securing Cyberspace for the 44th Presidency*, Washington, D.C.: Center for Strategic and International Studies, December 2008, p. 13, argues that better cyberdefense is required to prevail in today's constantly competitive economic environment:

> Ineffective cybersecurity and attacks on our informational infrastructure in an increasingly competitive international environment undercut U.S. strength and put the nation at risk.

> Our most dangerous opponents are the militaries and intelligence services of other nations. They are sophisticated, well resourced, and persistent. Their intentions are clear, and their successes are notable. Porous information systems have allowed our cyberspace opponents to remotely access and download critical military technologies and valuable intellectual property—designs, blueprints, and business processes—that cost billions of dollars to create. The immediate benefits gained by our opponents are less damaging, however, than is the long-term loss of U.S. economic competitiveness. We are not arming our competitors in cyberspace; we are providing them with the ideas and designs to arm themselves and achieve parity. America's power, status, and security in the world depend in good measure upon its economic strength; our lack of cybersecurity is steadily eroding this advantage.

[55] Marcia Savage, "Companies Still Not Reporting Attacks, FBI Director Says," *SearchSecurity.com*, February 15, 2006, reports Federal Bureau of Investigation (FBI) Director Mueller as saying that most companies that experience network intrusions do not report the incidents to law enforcement out of privacy and other concerns.

signal. Estimates of over $100 billion worth of annual damage in the United States alone are common.[56]

Nevertheless, certain things have definitely not happened in cyberspace. First, a long-standing annual survey of large organizations reveals that accounted-for costs have only recently exceeded $1 billion dollars.[57] Second, adversaries actively engaged against the United States (who thus have no reason to hold back for a more propitious time) have not conducted known cyber attacks; examples include Serbia in 1999, Iraq in 2003, and al Qaeda since at least 1998. Third, there have been no interruptions in power distribution in this country; the only interruption in phone service that can be traced to computer hacking took place ten years ago and involved fewer than 1,000 people.[58] But all that still leaves a great deal yet to be ascertained.

---

[56] Lawrence Wright, "The Spymaster," *The New Yorker,* January 21, 2008, p. 8 (online) says that "He [Mike McConnell, Director of National Intelligence] claimed that cyber-theft accounted for as much as a hundred billion dollars in annual losses to the American economy."

[57] New Media Institute, "2007 CSI/FBI Computer Crime and Security Survey," Web page, September 14, 2007. Such numbers should be extrapolated with considerable caution. It is hard to evaluate these results without knowing who did not answer the query and whether participants correctly assessed the costs of hacking. The survey also did not include DoD, perhaps the most assiduously targeted organization on the planet. RAND is carrying out a larger survey for the U.S. Bureau of Justice Statistics for 2009 release. See also Shane Harris, "The Cybercrime Wave," *National Journal,* February 7, 2009, pp. 22–29; the author must have been quite careful in not extrapolating too far from crimes of known cost because the cost of computer crime in the article never exceeds a billion dollars a year.

[58] In 1997, a teenager hacked into the Worchester, Massachusetts, communications system and disabled links to the local air traffic control computer (Pierre Thomas, "Teen Hacker Faces Federal Charges," *CNN.com,* March 18, 1998). The Blaster worm has been implicated in the August 2003 Northeast blackout, but its implication is incidental: The worm may have indirectly caused certain monitoring systems to be taken offline (Bruce Schneier, "Perspective: Internet Worms and Critical Infrastructure," *CNET News,* December 9, 2003; Dan Verton, "Blaster Worm Linked to Severity of Blackout," *Computerworld,* August 29, 200)3.

CHAPTER THREE

Why Cyberdeterrence Is Different

Cyberdeterrence seems like it would be a good idea. Game theory supports the belief that it might work. The nuclear standoff between the United States and the Soviet Union during the Cold War—which never went hot—provides the historical basis for believing cyberdeterrence should work.

It may well work. This chapter, however, lays out nine questions—three critical and six ancillary—that would differentiate cyberdeterrence from nuclear deterrence or general military deterrence. Such differences all work to the detriment of cyberdeterrence as a policy, and they illustrate why and how cyberdeterrence may be quite problematic. Expressed in terms of questions that are far less urgent when applied to, say, nuclear deterrence, the critical ones (*we* being the deterrer) are

- Do we know who did it?
- Can we hold their assets at risk?
- Can we do so repeatedly?

Six ancillary reasons are

- If retaliation does not deter, can it at least disarm?
- Will third parties join the fight?
- Does retaliation send the right message to our own side?
- Do we have a threshold for response?
- Can we avoid escalation?
- What if the attacker has little worth hitting?

This list, in toto, applies to the case of cyberretaliation following a cyberattack. Yet the relevance of individual elements is somewhat broader. Issues related to who did it, whether they can be disarmed, the right message, thresholds, and escalation apply to any retaliation following a cyberattack. Similarly, issues related to holding assets at risk, repeated retaliation, third parties, and escalation apply to cyberretaliation in response to other kinds of attacks. The little-worth-hitting argument applies primarily to cyberdeterrence in kind. This is illustrated in Figure 3.1.

The contrast with Cold War–era nuclear deterrence is obvious, but here it is, anyway. In a nuclear war, who did it is usually clear,[1] and targets can be held at risk. Attacks can be continued as long as weapons

Figure 3.1
Where Each Caveat Applies

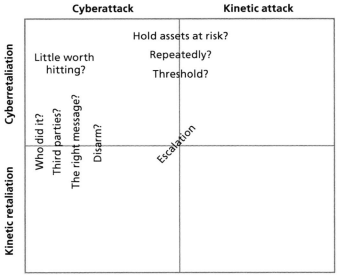

RAND MG877-3.1

---

[1]   Or at least was in the days when the United States had only one nuclear-armed opponent. Even now, the only way that identity would be in question would be if the vehicle were something like a suitcase weapon. Yet the forensics on the radio nucleotides left behind would offer pretty good hints.

and delivery vehicles survive. Nuclear storage and delivery infrastruc-tures can be disabled by nuclear attacks—which is precisely the role of counterforce targeting. The involvement of a third-party nuclear-armed state, much less nonstate nuclear warriors during an exchange, is highly unlikely. Private fortifications are of limited use (no one ever pretended that fallout shelters could prevent all damage). The consen-sus is pretty strong that any nuclear use (with its telltale residues) would be clear and would cross a major threshold. Nuclear warfare trumps all other forms. All states had their survival at stake. Very similar state-ments can also be made about conventional deterrence, such as using the threat of strategic bombing to inhibit land-based aggression.

The rest of this chapter delves into these nine differences indi-vidually and concludes with a brief discussion of how the will to retaliate, always a factor in the nuclear standoff, is less of an issue for cyberretaliation.

Do We Know Who Did It?

The notion that the one should know who attacked before retaliating seems clear enough. If deterrence is to work before the first retaliation takes place, others must have confidence that the deterring state will know who attacked it. Hitting the wrong person back not only weak-ens the logic of deterrence (if innocence does not matter, why be inno-cent?) but makes a new enemy. Instead of facing one potential cyber-war (against the original attacker), the defender may now face two (the second against the one incorrectly identified as the original attacker).

The value of attribution, and hence its difficulties, go deeper than that. The defender must not only convince itself but should also con-vince third parties that the attribution is correct (unless retaliation is kept quiet, and only the victim of retaliation can tell that is has taken place). Finally, and most importantly, the attacker has to be convinced that the attribution is correct. If the attacker believes the retaliator is just guessing or that the retaliator has ulterior motives for retaliating, it may conclude that carrying out further attacks will have no effect on whether or not it will face further punishment.

The need to convince third parties that an attribution is correct depends on the importance of third parties. In contrast to the bilateral nuclear standoff of the Cold War, third parties matter these days; over 100 countries are supposedly developing what are described as cyberattack capabilities (many may just be CNE).[2] There is not even a *single* dominant threat; the most putatively capable threats (e.g., China) are not the most hostile to the United States. Unlike the attacker, which is likely to know that it attacked (but not necessarily whether it did so successfully), third parties may not even be convinced that the retaliator was really attacked or had struck back for unrelated reasons. If the purpose of an attack was to corrupt a target system, effects will be apparent only to the attacker and the target (if even then). Even if the effects are public, the cause of the malfunction may be apparent only to the target (if correct) and the attacker (who will likely correlate the failure of the target system with its having been attacked). If retaliation is to be public, deterrence must likewise be public.

The last stricture seems funny: How would the attacking state know that it was *not* responsible? Most of the time (see Chapter Four for exceptions), the attacker would know. But the point is that the attacker has to believe that the target knows and that the target is retaliating because it was attacked. The difficulty of attributing (and even detecting) all cyberattacks makes this less than obvious. For instance, the attacker may know that it carried out an attack, and in fact had been carrying them out for years, but that other states had as well. The attacker would thus ask two questions after retaliation: Why me? Why now?[3] What the attacker concludes may have little to do with the fact,

---

[2]   John Swartz, "Chinese Hackers Seek U.S. access," *USAToday*, March 12, 2007, spoke with Jody Westby, CEO of Global Cyber Risk: "'The Internet was not designed for security, and there are 243 countries connected to the Internet,' says Westby, who estimates 100 countries are planning infowar capabilities." The article from which the quote comes discussed Chinese CNE against U.S. military computer systems.

[3]   In case of a delay, the reason might be that retaliator has just issued a "new" deterrence policy. The retaliator can announce such a policy after it has finally been able to attribute an attack but will pretend that the attribution actually followed the announcement. Third parties might be fooled, but the attacker—who may well know that the attack preceded the announcement—may be more suspicious. Fortunately for the potential retaliator, the

much less the particulars, of the cited attack. Instead, it may reflect other events—a bureaucratic tussle within the retaliating state, a nasty trade dispute, an attempt to win a third state's favor. So thinking, the attacker may decide to halt attacks anyway and give the retaliator no further excuse. Or it may figure that further attacks are unlikely to raise the odds that it will be subject to retaliation, especially if the retaliator offers no evidence that implicated the attacker.

Attribution may be so uncertain that the odds that any one cyberattack could evoke a response would be fairly low. How low can the odds of attribution fall without destroying the empirical basis for deterrence? The raw calculus of deterrence is fairly straightforward: The lower the odds of getting caught, the higher the penalty required to convince potential attackers that what they might achieve is not worth the cost. Unfortunately, the higher the penalty for any one cyberattack, the greater the odds that the punishment will be viewed as disproportionate—at least by third parties (who will not know what the attacker did get away with) and perhaps even by the attacker. In other domains with low catch rates (e.g., traffic violations, marijuana possession), the accused at least know that they were caught because they were guilty.

What makes attribution so hard? In a medium where "nobody knows you're a dog,"[4] it is equally hard to know whether you are a hacker. Computers do not leave distinct physical evidence behind.[5] The world contains billions of nearly identical machines capable of sending nearly identical packets. Attacks can come from anywhere.[6] State-

---

attacker may be less than eager to prove when it had attacked, lest it admit that it had, in fact, attacked.

[4] Peter Steiner, cartoon, *The New Yorker,* Vol. 69, No. 20, July 5, 1993, p. 61.

[5] Internet Protocol (IP) version 6 (IPv6) permits better attribution than IPv4 because it tracks the source of a packet more reliably but far from well enough to put a real dent in the attribution problem. One might imagine a future in which Internet packets are reengineered to show some machine-specific identification number, but do not expect it very soon. IPv6, the urgent need for which was recognized by the early 1990s, well before the world ran out of address spaces, has yet to be completely implemented, particularly in the United States. Even when it is, the first few versions of any attribution-friendly IP would hardly be spoof proof.

[6] David A. Wheeler and Gregory N. Larsen, "Techniques for Cyber Attack Attribution," paper, Alexandria, Va.: Institute for Defense Analyses, October 2003, presents 17 differ-

sponsored hackers could operate from a cybercafé, a public library with Wi-Fi access,[7] or a cutout.[8] Finding rogue packets that can be traced back to the network (IP) address of a government bureaucracy reveals a bureaucracy that is stupid, is arrogant, runs so many hackers that it cannot be anything less than obvious, or operates a network that has been hijacked by others. Packets can be bounced through multiple machines on their way to the target. They can be routed through a bot that only needs to erase the packet's originating address and substitute its own to mask the true origin. Attacks can be implanted beforehand in any machine that has been compromised.

The test that presumes that the beneficiary of action was its most likely instigator (cui bono) can be misleading. If the attack appears to have been made on behalf of a cause (e.g., Palestinian rights), one of several states—or none of them at all—may be behind it. A greater risk is the possibility of false-flag operations designed to get another state in trouble. Indeed, the more serious the threat of retaliation, the greater the incentive for false-flag operations on the part of the presumed attacker's enemies. A state's failure to cooperate with the investigation of a particular incident may be telling (and, as discussed later, can be made more telling) and may thus be construed as an indication of guilt—or nothing more than evidence that someone has some other state secrets to protect. Even friends whose cooperation may be needed

---

ent attribution techniques, but these require very high levels of cooperation among router owners worldwide and reveal only which machine the attack packets are coming from (which may be a bot and hence point only to a slob, not to the attacker). Many can be foiled easily if their use is anticipated. Finally, as noted, the correlation between machine and person can be quite low.

7   Consider this from Peter Walker, "American Expats Caught Up in Indian Bomb Blast Inquiry," *Guardian.co.uk*, July 29, 2008:

> When Indian police investigating bomb blasts which killed 42 people traced an email claiming responsibility to a Mumbai apartment, they ordered an immediate raid.

> But at the address, rather than seizing militants from the Islamist group which said it carried out the attack, they found a group of puzzled American expats.

> In a cautionary tale for those still lax with their wireless internet security, police believe the email about the explosions on Saturday in the west Indian city of Ahmedabad was sent after someone hijacked the network belonging to one of the Americans. . . .

8   In intelligence terminology, a *cutout* is someone who operates on behalf of another.

to trace packets back to their source may hesitate if they think successful attribution will lead to a crisis. Furthermore, rejection may be entirely innocent; U.S. (or European) courts, for instance, could reject some investigative techniques other states employ because they violate privacy rights.[9]

The occasional difficulty in determining whether a system's failure did, in fact, result from a cyberattack further complicates attribution. If events in the real world are indicative, the popular tendency will be to assume that any spectacular computer failure resulted from hostile action. Note how often bystanders are quoted as thinking they were under terrorist attack when they heard a loud noise (e.g., a collapsing crane,[10] a natural gas explosion). For several hours on November 12, 2001, New York City reacted to the crash of American Airlines 587 as though terrorists had done it again.[11] Cyberspace has no fewer alternative explanations. When a system goes haywire, it could be bad software (which has explained many widespread outages), human error, or natural accidents (the Northeast power outage in 2003 can be traced back to untended trees in Ohio).[12] Even deliberate attacks may have a "whoops" component to them: The infamous Morris worm of 1988 was supposed to spread too slowly to affect the Internet, but one parameter in the code was set incorrectly.

When attribution can be localized to a country, or even to government networks, that fact does not in itself prove that the attack

---

[9] In some countries, what investigators are looking for may not be illegal. In Argentina, a group calling themselves the X-Team hacked into the Web site of that country's supreme court in April 2002. The trial judge stated that the law in his country covers crime against people, things, and animals but not Web sites. The group on trial was declared not guilty of breaking into the site (Paul Hillbeck, "Argentine Judge Rules in Favor of Computer Hackers," *SiliconValley.com*, February 5, 2002).

[10] "We're being bombed," was one reaction from a witness to the collapse of a crane in Manhattan (Manny Fernandez, "Terrible Rumble, Then Chaos as Crane Fell," *New York Times*, March 16, 2008).

[11] David Johnston and James Risen, "The Crash of Flight 587: The Investigation," *New York Times*, November 13, 2001, p. D9.

[12] New York Independent System Operator, *Final Report on the August 14, 2003 Blackout*, Rensselaer, N.Y., February 2005.

came from that state, that is, from someone operating under national command. It could be. But it could also be an element in the government, perhaps operating on behalf of what it perceives to be state interests but without specific or at least not clear authorization (it may have permission to spy but not to tamper). Or the attacker could believe its activities were winked at by a government that wanted to preserve deniability. Or the attacker could be a proactive bureaucratic faction that acted when it deemed command authority wimpy. Or an attacker could be an entrepreneurial group of hackers who were looking to steal information (if CNE) or to create effects that it was confident would be appreciated and perhaps even rewarded—in outright cash, by future contracts, or by having its other illicit activities overlooked.[13] A large proportion of all those with DoD network addresses are actually support contractors, creating the (admittedly largely theoretical) possibility that these individuals are answering to their employers, not the government (if they are cutouts, the government is responsible). Hackers may be off the government payroll but linked to a particular political faction or to individual politicians (more likely in non-Western states). They may want to further state interests as their friends perceive them or may want to get the current regime in trouble—the better for their friends to assume (more) power. The hackers may be organized criminals (e.g., the Russian mafiya noted above) who have co-opted the state. The hackers could be "superpatriots" who have no connection to the government or ruling elites but are striking at adversaries in lieu of or in advance of where they are sure the government would go.

---

[13] If the account in Grow, Epstein, and Tschang, 2008, is true, the hackers, supposedly Chinese, who sent a rogue email to a vice president of Booz Allen Hamilton also sent a blind copy to James Mulvenon, whose counterhacker activities are apparently well known in China. One has to believe that such a look-what-we-can-do side message is not the work of national security professionals but amateurs, albeit talented ones. Although it is plausible to imagine the government taking advantage of opportunities freelancers provide, it is implausible to imagine it using freelancers when it has similar cyberattack resources under its command. Paying freelancers offers only a slight advantage in after-the-fact deniability. But the government has less control over freelancers than it has over its own staff. The former may use amateurish techniques; may wander from the designated target list; and, worse, may be in the target's pocket and thus eager to implicate when "caught."

The following hints may be indicative. Private hackers are more likely to use techniques that have been circulating throughout the hacker community. While it is not impossible that they have managed to generate a novel exploit to take advantage of a hitherto unknown vulnerability, they are unlikely to have more than one. In contrast, state hackers can tap a larger and more-secretive research effort that can consolidate discoveries, tools, and techniques across their own organization. Thus, state hackers are more likely to have a sizable bag of novel tricks. The approaches of state hackers, especially if operating against multiple sites, would tend to be more methodical and uniform (as befits a uniformed military, for instance) and less likely to be experimental or whimsical.[14] Style points matter little to states but matter greatly to hackers eager to impress their friends. State hackers are likely to provide more-consistent and even coverage through the day, patiently waiting until people are least likely to be standing guard over their systems. Such hackers are more likely to have the resources for round-the-clock coverage (although freelance hackers may occasionally provide team coverage, each hacker's individual style may show through from one hour to the next). Similarly, state hackers are likely to be more disciplined in attacking certain targets for certain reasons and avoiding others that may look equally interesting but are not part of the plan. Freelance hackers, by contrast, are more likely to explore randomly, indulging in the thrill of discovery. Perhaps most tellingly, only state-sponsored hackers are likely to go in with substantial knowledge of the target's military operational systems (e.g., how surface-to-air missiles [SAMs] fail) rather than simply its military information systems (e.g., how SAM computers fail). To cause the desired effects, state-sponsored hackers have to understand the machinery they are trying to interfere with. If such knowledge is classified, it is likely to be possessed only by someone with a security clearance: a government official or a closely monitored contractor. For example, evidence that an attack used heavyweight code-breaking would normally point to a

---

[14] For a popular account of Titan Rain, an effort to counter the Chinese cyberspies, see Nathan Thornburgh, "The Invasion of the Chinese Cyberspies (and the Man Who Tried to Stop Them)," *Time*, August 29, 2005b.

state attacker because traditionally, private hackers could not afford the supercomputers necessary for such a task. Still, even this may no longer differentiate the two as well as it used to. In 1999, RSA Data Security announced that a long-standing encryption standard had been broken using the brute force of networked computers.[15] A hacker or hacking group that can rent a botnet whose bots are assigned to code-breaking may be as effective as most states.

Finally a third class of hackers, those organized and financed by criminal enterprises, are likely to have some attributes of both state and freelance hackers but can be differentiated according to their targets and aims. Nevertheless, distinctions between states and freelance hackers (and/or criminal hackers) are probabilistic and are not based on a great deal of revealed real-world experience.

None of this is to say that attribution is necessarily impossible. Attackers may be stupid (e.g., in operating from an address linked to the state). They may be arrogant and thus sloppy. They may be brazen and not care whether they drop hints, because hints are not proof, or because they simply do not care. Open chatter may prove a state's unmaking, especially if it uses hackers who are currently freelancing or have previously freelanced and then talked about what they did. Finally, states may be penetrated.

A state that has carried out CNE in high volume may also reveal a modus operandi (MO), which can be used to trace the source of an attack. The high volume involved in CNE (a great deal of hay must be moved to find the occasional needle) facilitates extraction of MOs. Operating at high volume leads to pattern repetition, use of automation (such as bots), or enlistment of an army of less-sophisticated (albeit thoroughly trained and well-disciplined) hackers using similar methods.[16] Mere replication of methods may suggest that such an army is

---

[15] RSA Data Security, "RSA Code-Breaking Contest Again Won by Distributed.Net and Electronic Frontier Foundation (EFF)," press release, San Jose, Calif., January 19, 1999.

[16] How can one distinguish a replicable bot from a disciplined cadre of humans? The latter may have to show enough human insight in its approach to pass a Turing test. Telling evidence of human behavior may emerge if the site mandates passing an I-am-human-and-not-a-bot test (e.g., identifying bizarre shapes as letters and numbers). Minor differences in approach between one attack and the next may also indicate the human touch.

at work; if so, this conclusion would limit the search to states large enough to employ the requisite manpower. CNE also requires a "to:" address on exfiltrated information; the sheer quantity of material can help investigators trace packets back to their ultimate destination (until peer-to-peer distribution via bots becomes the norm). Once the MO is established, it may be possible to find something similar when analyzing cyberattacks. Conversely, an attacker that believes it is important to preserve anonymity may isolate its cyberattackers from its cyberspies.

Nevertheless, in contrast to the increasing transparency of the physical world, there is no ipso facto basis for believing that attribution will be better tomorrow than it is today. Indeed, cyberspace is, if anything, growing increasingly noisy—in large part because computers and networks are becoming more complex, making it easier to hide tell-tale signals.

As hard as attribution is today, when a state caught with its virtual hand in the virtual cookie jar faces few penalties, it would likely be much harder if attackers faced retribution if caught. Deterrence may inhibit states from attacking in the first place, but it is just as likely to persuade them to cover their tracks more carefully and continue attacking. After all, as already noted, they have many ways to do so, including working from overseas and avoiding tools, techniques, and hackers with which they have already been associated.

Incidentally, recovery may suffer if the threat of retaliation persuades attackers to hide better. Attribution permits diagnostic forensics (insofar as specific states have signature MOs), which, in turn, helps reveal the source of the damage and thus may hint at how to reverse it. Anything that persuades the attacker to erase evidence that may lead to attribution also tends to reduce the amount of information that defenders can use in system repair.

Even if an attribution is correct, the challenge remains to convince the attacker that it has, in fact, been caught doing something specific and has not simply been fingered because it was unloved by the target. The most direct method—"here is the evidence that you did it"—may convince third parties but may simply not be good enough for the attacker.[17] Worse, the evidence may reveal forensic methods or aspects of the target system that are better kept secret—especially if showing the first set of evidence creates demands for backup evidence.

An overly clever way to communicate the "gotcha" message (while avoiding the demand for backup information) would be to retaliate using the same "signature" MO (if it can be discerned and duplicated) that the attacker used, thereby declaring, first, that the retaliator knows that it had been attacked and with what MO and, second, that the retaliator associates this particular MO with that specific attacker. But, satisfaction aside, success requires the attacker's leadership to understand that the retaliator's MO is, in fact, the same as its own MO. This brings up a recurring difficulty of setting rules in cyberspace: Too much of it is simply too complex for decisionmakers to understand. They would have to fall back on experts and trust that the experts are not defending their backsides by denying the similarity (admitting the similarity is admitting that one has been caught). Conversely, the returned MO may simply fail. The attacker, aware of the vulnerability the MO exploits, may have closed the vulnerability in its own systems; the retaliation would then fail; and the attacker (as defender against retaliation) never knows the attempt was made. Alternatively, the attacker may believe that the MO is too obvious to "belong" to any one attacker.

One saving grace is that the attacking state may simply reveal itself. Revelation can be private (to the target) or public. Private revelation can be carried out through diplomatic or intelligence channels or by leaving a "calling card" in cyberspace (e.g., information that only the attacking state would have). Public revelation would probably have to come up with evidence that gave such announcement credibility. Despite the many reasons revelation is ill-advised, it may be justified if the aim is coercion. Perhaps the attacker wanted to be clear about what the target is supposed to do or not to do to escape future episodes. It may feel that implicit coercion may not send a sufficiently clear mes-

---

[17] Evidence of one attack does not necessary address the why-now question. Trying to prove that the retaliator simply had no good evidence on prior attacks by the attacker or by others is nearly impossible.

sage. Absent explicitness, the target may have implicitly blamed someone else or may simply have failed to recognize that what it perceived as
system failure was an actual coercive attack. If the effect of the attack
is not obvious, the attacker needs to know that the target knows it has
been hit, lest the attacker appear silly in taking credit for something
that does not seem to have taken place. Finally, the attacker may want
to get the attention of third parties and believes the target will either
not publicly identify the source or that third parties will not credit
the target's evidence and thereby not believe the identification. The
attacker may also go public because it believes that private "calling
cards" may not get passed up to the target state's command authorities.
Confession, at least, should end the problem of attribution—unless the
confessor is lying, which it might do to take credit for something someone else did (rather daring, unless it is sure that the real hacker will
not contradict such claims) or is protecting another, perhaps weaker,
state more likely to be subject to retaliation than the confessor state
would. Many examples of misattribution can be drawn from the world
of crime and terrorism; note that Khalil Sheik Muhammad confessed
to more terrorist incidents than he was entitled to take credit for.[18]

One final consideration. If you cannot tell who did it or even
communicate what the damage was, you also cannot tell who did not
do it or what the damage could have been. As long as the burden of
proof is not heavy or, better yet, if it can be shifted to the accused, who
has to prove the negative, the supposed target can claim an attack that
may not have happened from someone who probably did not do it. But
anyone who goes down that road is probably not interested in the cal-

---

[18] Adam Zagorin, "Can KSM's Confession Be Believed?" *Time*, March 15, 2007, reported
the following:

> [Khalid Sheikh Mohammed] admitted under previous interrogation that a list of 30
> supposed U.S. targets, which he circulated shortly after 9/11, was a lie to exaggerate
> the scale of al-Qaeda's planning. Terrorism experts say that though there is no doubt
> Mohammed played a major role in planning 9/11, he's famous among interrogators for
> his braggadocio. "He has nothing else in life but to be remembered as a famous terror-
> ist," says Bruce Riedel, Senior Fellow at the Saban Center at the Brookings Institute and
> a 29-year veteran of the Central Intelligence Agency. "He wants to promote his own
> importance. It's been a problem since he was captured," says Riedel, who went on to say
> he wouldn't be surprised if Mohammed was exaggerating his role in other plots.

culus of deterrence and is more interested in justifying aggression and sowing mischief and just needs some rhetorical cover. Such purposes may be incompatible with U.S. behavior, but establishing or at least supporting norms that legitimize cyberdeterrence may give less fastidious governments yet one more excuse to wreak international mischief.

Can We Hold Their Assets at Risk?

Battle damage is a multifaceted issue. Before the attack (or retaliation) is launched, the attacker does not have a good idea what the sum of its effects will be—and the target does not know what the attacker is capable of damaging. Even afterward, neither the attacker nor even the target may know for sure what the damage was. It is one thing to assess an attack that blows up a refinery and thereby eliminates a source of gasoline; it is another to assess an attack that corrupts the refinery control system to introduce subtle but vehicle-damaging changes to the chemical mix in the gasoline.

Battle damage prediction is critical in establishing deterrence at all. All deterrence requires the ability to hold something at risk. Yet without knowing which targets are vulnerable to what degree—and, more unpredictably, how quickly they can be recovered—it is difficult to know, much less promise, what damage retaliation can wreak. From the retaliator's point of view, the worst outcome would be to huff and puff after the attack, announce that retaliation would follow, carry it out—*and no one notices.* Waiting too long and claiming success after the one act of retaliation that manages to succeed enough to catch the attacker's attention creates other ambiguities: Was the return strike a retaliatory blow, or will it be perceived as aggression and thus start a new cycle? Without a declaration, is it certain that the retaliator carried out a delayed strike, or did the retaliator claim success for an attack that some third party actually carried out for other reasons? After all, no single target in this world has just one potential attacker. This is why claiming to put any specific target at risk from a cyberattack is foolish: The more specific the asset put at risk (e.g., China's Three Gorges' Dam in exchange for messing with the Hoover Dam), the more likely that asset will receive additional protection (or merely be yanked offline) and therefore be placed out of risk.

Even if the effects of retaliation on any one target are unpredictable, might not multiple attacks produce some statistical level of certainty? They might, were not failures often correlated with one another (e.g., the target has unseen defenses against a particular class of attack). Beyond that, difficulties in predicting human performance (security sensitivity, damage mitigation, repair) bedevil even order-of-magnitude guesses. Also bear in mind that the point is not to create just *any* effect by retaliation but enough effects to catch enough of the attention of policymakers that they begin to calculate that the costs of carrying out further attacks would not exceed whatever benefits they might gain.

For this reason, it is unclear how policymakers can gauge the effectiveness of a contemplated retaliation, a prerequisite for a deterrence policy. They may remember that, from World War II to Vietnam and onward, strategic targeteers have overstated how long enemy infrastructures would be unavailable if destroyed—and cyberspace is a far more difficult environment to make such calculations for.[19]

Potential retaliators also face the prospect that, without a true understanding of which downstream computer processes depend on the targeted system or software, a retaliatory attack will cripple or corrupt operations well beyond those intended. This may create problems for the retaliator. If the damage is disproportional, others may condemn it as such. The defender may conclude that it is escalatory and respond with counterescalation. The retaliator may also have lost the opportunity to declare certain classes of targets off limits in hopes that the attacker may observe similar thresholds.

For either attacker or retaliator, good battle damage assessment (BDA) requires answering many questions correctly: Was the target penetrated? Did the attack affect the functioning of the target (and is the damage real or feigned)? If the system supports human decisionmaking (or if its malfunctioning would be important to decisionmakers), were

---

[19] For a broader survey, see T. W. Beagle, *Effects-Based Targeting: Another Empty Promise?* thesis, Maxwell AFB, Ala.: School of Advanced Airpower Studies, Air University, June 2000.

the effects noticed—and by the decisionmakers? If the intent was to coerce, how do we know it was persuasive? Negative answers here show that an attack can succeed in technical terms and fail to register in operational terms. Has collateral damage been minimized or at least accounted for? After all, the target of a cyberattack is a system that was not supposed to be easy for random hackers to get into (otherwise it would have been hit already). If the vector of the attack had any self-replicating code, access to the target system may be yanked (even literally, if a network wire is pulled) after the code was inserted but well before the full damage has taken place. Absent a monitor resident in the target system, the only way an attacker might know what happened is if the attack was designed to disrupt a service available to the public, e.g., the lights go out. But such targets are not necessarily the best ones in terms of the kind of damage they can cause.

This dilemma holds with even greater force for retaliation. Since its *primary* purpose is to communicate displeasure, it may be even more necessary to make only the attacks that produce obvious effects— even though they are not necessarily the most damaging attacks to the victim (the original attacker) and may be easier to reverse than the more-subtle attacks are. Otherwise, it may not be known whether the message has been received on the other end.

One might think that the target, at least, knows what the damage has been. This will be mostly true for disruption—but only if the target knows that the disruption was caused by an attack rather than some malfunction. This may not be always be true for corruption—the point, after all, is to ruin processes in ways that defy detection and correction but are not immediately recognized as such (ruining processes in ways that obviously require restoration can only be second best from the attacker's perspective). Ironically, the attacker may have a better fix on what happened because it knows which systems or processes were being targeted, while the target can only guess.[20] But the attacker may have no good sense of which systems or processes relied on the corrupted systems or processes; only the target will know that.

---

[20] This holds even more strongly for CNE. The attacker will know what the take was, while the target usually has little clue.

The prospects for improving battle damage prediction by building and operating test ranges where attacks can be run on mock-ups of target systems may prove largely illusory. Such test ranges exist.[21] But repeating the same attack on instrumented test ranges (where all parameters can be prespecified) can yield widely varying results. Internal computer processes of one sort or another (e.g., self-examination, polling software threads to see what needs attention, context switching between tasks) are always taking place. It is not unusual for an attack to elicit a weak response if the computer happens to be in a state subtly different from the one it had when the attack worked. The ability to know all a system's parameters beforehand is hardly guaranteed— and this is for a relatively short "beforehand" that may apply over the planning of an attack. No software-based system will stand still in the months and years between the decision to incorporate that system in a state's retaliatory target list and when the system is actually struck.

Discovering a specific vulnerability does not mean it will be there when the time comes to exploit it. The target may have discovered the exploit but reacted in ways that are undetectable to the attacker. Military or intelligence operators may well program around the vulnerability so that it generates what, to attackers, may be indicators of a failure that never, in fact, happened. Finally, the adversary itself may not know how it might react to an attack; its warfighters may, under stress, evidence an operational agility that the adversary would not have been able to predict—or they may seize up altogether.

Testing the attack in vivo without revealing its existence takes nerve. Examining the target computer's file structure can reveal whether altered code or files are still there (whether operating programs are still referring to them is another matter). The brave can execute attacks up to just short of the point at which they would create effects of the sort that the target's defenders would indeed notice—meaning that, supposedly, only the attackers would notice the effects of such attacks. If the attacker uses implants, it can query them to see if they respond to

---

[21] The National Cyber Range has been estimated as having an ultimate price tag of $30 billion (a number that seems quite high). See Noah Schiffman, "DARPA Attempting the Impossible: Self-Simulation for Defense Training," blog, *Network World*, June 6, 2008.

test signals correctly (a good hint that they will respond to real signals correctly). Too much testing, though, could alert defenders and cause them to find and patch the vulnerabilities that permitted such attacks (the in vivo tests) in the first place. Even successful in vivo testing leaves open the question of how dynamic the target sysadmins' detection and repair practices are.

The graver difficulty is reconciling the difference between in vivo testing and in vivo results. Only in vivo testing will be reliable because the efficacy of any attack is specific to the defenses or recovery practices associated with the target network. Some of the defenses, such as intrusion detection or packet inspection systems, may be physically outside the target system (see the discussion of government-provided defense in Chapter Five). These defenses may escape the attacker's notice.

Can We Do So Repeatedly?

Deterrence can be fragile if hitting back today prevents hitting back tomorrow and thereafter. For most forms of deterrence, this is not a problem. Some deterrents are so awful that no one tempts them. For others, one hit does not preclude another. In cyberspace, the problem is vexing. Serial reapplication of retaliation may be necessary, but each use tends to diminish the expected consequences of the next use.

As previously discussed, the ability to penetrate a system and make it do what it was not designed to do requires the target system to have a vulnerability. If the system is attacked and if the attack is recognized as such (rather than, say, dismissed as a normal glitch), sysadmins will understand they have a vulnerability of some sort. If the vulnerability is known but is not attended to, sysadmins will likely hasten to catch up on their repairs (e.g., installing the requisite patch, closing the offending port). If a fresh vulnerability is discovered, efforts will be made to repair the vulnerability directly or, if the software came from elsewhere, tell the software vendor and press for a solution. If the fact but not the nature of the vulnerability is known, sysadmins may route around the offending system or code (e.g., by disallowing functions or settings that activate the code). Granted, success is not assured.

The attack may not be correctly identified. The discovered vulnerability may be a manifestation of a deeper problem that goes uncorrected. The fix may break something else or open up new vulnerabilities. But repetition is not assured, either.

If the attack works by looking for vulnerabilities in the periphery—e.g., user-managed systems—there is little guarantee that the same human weaknesses (e.g., clicking on a rogue Web site, yielding password information to tricksters) will not lead to subsequent problems ad infinitum. Whether or not peripheral vulnerabilities are consequential and, if so, how consequential, is another question.

As a general rule, tricks exhaust themselves to the extent (1) that their existence and thus the need to protect against their recurrence is obvious and (2) that counters to their recurrence are straightforward to implement. Certain types of attacks will deplete at different rates. The depletion rate for computer network exploitation (by way of comparison) is relatively low because it often goes unnoticed for long periods. One would expect the depletion rate for an obvious disruption attack to be fairly high, but such an attack tends to be easier to implement in the first place because it requires introducing error into a system. For corruption, the opposite is true: Depletion is fairly high (*fairly* because corruption may go unnoticed and hence unaddressed but *high* because attempts to carry out a new corruption attack are difficult). Success requires subtly and consistently altering many parts of a system, lest telltale inconsistencies alert defenders that something is amiss.[22]

Overall, it is best not to count on the same trick working forever or even for very long. It is, for example, almost unheard of for a dangerous virus or worm to run amok in well-defended systems after it has drawn so much attention to itself that antidotes are developed and circulated.[23] Even variants of such malware tend to have far less effect

---

[22] If there are, for instance, two different programs—one to record additions and withdrawals and another to inventory stocks—corrupting one without corrupting the other in the same way will yield a telltale contradiction between the two readings.

[23] Early, less-efficient versions of a virus may yield to later, more-efficient versions of the virus if the virus maker learns from its shortfalls faster than defenders learn about the virus and take actions to ward it off. The original Sobig.A version, discovered in January 2003, went through a series of revisions; it took another seven months for the most successful

than the original.[24] Thus, a trick taken out of a bag cannot always be put back for use another day. Conversely, since vulnerabilities are constantly being discovered and corrected, the half-life of an exploit may be limited, leading to a bit of a use-it-or-lose-it dilemma.

Depleting the inventory of potential cyberattacks means that discovering new vulnerabilities will require more effort and will take longer. Creating exploits for whatever vulnerabilities are subsequently discovered may require more effort, with each additional step creating room for error. Even after being set up, the later exploits may (1) have less-damaging effects, (2) require riskier actions (e.g., social engineering, on-site access, insiders) to emplace, or (3) work only in a smaller set of circumstances (e.g., while a user has a computer in configuration A and undertakes task B). Some level of repletion is possible through intensive search, technical means (e.g., research and development, modeling, probing, and decompiling), and human intelligence on the target's designers. The target's actions may also help: Sysadmins may turn over, and new ones may not remember what such an attack looks like—or old ones may forget or grow complacent. In any case, because systems change—in large part when new vulnerabilities are found—no one can sit on an unused bag of tricks for very long.

The difficulty of continuing attacks (whether original or retaliatory) complicates predicting what a follow-on retaliation might achieve. If the retaliator is to threaten a similar attack the next time, it may know what the first one achieved but can only guess how the victim fixed (or routed around) its hitherto-vulnerable systems in response to the first retaliation. In contrast to most forms of warfare, repeated use does not necessarily improve anyone's understanding of weapon effects.

This problem has direct ramifications for the attacker's behavior. Even if the initial retaliation was painful, the attacker may be convinced that its fix was enough to safeguard it sufficiently. So the attacker con-

---

of these, Sobig.F, to appear (Thomas M. Chen and Jean-Marc Robert, "The Evolution of Viruses and Worms," in William W. S. Chen, ed., *Statistical Methods in Computer Security*, CRC Press, 2004.

[24] Since the mid-1990s, vulnerabilities have migrated from viruses to macros to worms (infecting "servers") and from operating systems to applications as the earlier opportunities for mischief have been recognized and reduced.

tinues its mischief.[25] Thus, a second retaliatory attack may be required. Even then, the attacker may be convinced that that the second fix worked. Proving to a stubborn attacker that diminishing returns are not setting in may require multiple attacks—if possible. Conversely, all this may be an academic quibble: Despite professional optimism that the original vulnerability has been laid to rest, the public may believe otherwise, and it is the public's reaction that may shape the attacking state's behavior.

All this depletion and fragility applies to the attacker as well. That is, the undeterred attacker will find it continually harder to hit similar targets because they harden as they recover from each new attack. It is possible to argue that, while the quality of retaliation is depleting, so too is the quality of the attack that retaliation was meant to deter. From another perspective, this means that the importance of deterrence vis-à-vis defense is likely to decline with repeated use.

Figure 3.2 illustrates, for the sake of clarity, the point that the efficacy of cyberattacks as a function of how frequently they are repeated is different from the efficacy of cyberattacks as a function of intensity (see Chapter Two), and both are different from the efficacy of cyberattacks over time.

If Retaliation Does Not Deter, Can It at Least Disarm?

Retaliation attacks are useful only for deterrence. Unlike conventional or nuclear retaliation attacks, they are generally incapable of disarming the attacker. If retaliation does not build deterrence, there is no second prize here. This is why.

The prerequisites for a cyberattack are few: talented hackers, intelligence on the target, exploits to match the vulnerabilities found through such intelligence, a personal computer or any comparable computing device, and any network connection. Powerful hardware may be

---

[25] Why the does attacker not run out of tricks itself? The answer may well be that its inventory of tricks also declines, but the retaliator cannot necessarily wait for such declines to set in (and new software with new bugs may present new opportunities for mischief).

Figure 3.2
Three Dimensions of the Efficacy of Cyberattacks

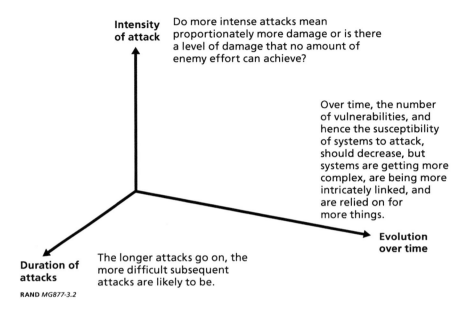

RAND *MG877-3.2*

needed for breaking codes or decompiling software, but it need not be online. If that hardware is not online, it is very difficult to break with a cyberattack. Botnet attacks are certainly useful, albeit less sophisticated, but botnets can be rented and cannot be destroyed through cyberattack, almost by definition.[26] No botnets are easy to dent, much less destroy, through cyber counterattack. Indeed, since hackers need only an arbitrary computer and one network connection, it is not clear that even a physical attack could destroy a state's cyberattack capabilities (unless their hackers cluster in one physical location).

---

[26] Within the Linux community, someone came up with a "friendly" worm whose function was to search the Internet and destroy a known "unfriendly" worm (Bryan Barber, "Cheese Worm: Pros and Cons of a 'Friendly' Worm," SANS Institute, 2001). His efforts were not applauded. While it would not be totally impossible for someone to invent a virus that hunts down and disables bots, it would debatable whether the unsuspecting users whose machines had hosted the bots would consider this to be an attack in and of itself.

That being so, there are serious drawbacks to an active defense—defined as automatic targeting of the attacking computers.[27] Perhaps it is satisfying to stop an attacking machine while an attack is in progress or to see if one can capture attack tools from such a machine.[28] But an active defense may also automate retaliation decisions that could profit from more-careful consideration. An attacker who anticipates active defense can easily make the attack appear to come from somewhere else, ranging from the sensitive (e.g., an orphanage, mosque), to the ticklish (e.g., an opposition newspaper, a trusted ally), to a computer within the target system itself. The attacker could also establish a honey pot in front of the attacking computer to capture the return packets and thereby analyze the target's retaliatory capabilities. Some types of attacks, notably those involving bots, do not necessarily have a single point of origin. Finally, even if the attacker's computer were destroyed, the attacker would be out only a few hundred dollars.

Combining the last two tenets suggests that a state's cyberattack capability is more likely to lose its punch by being used than by being attacked.

The inability to disarm attackers has three silver linings. First, the inability to destroy a cyberattack capability means that preemption is not a rational motive for attacking others; hence, this is one less reason to start a fight. Second, the attacker can see retaliation for what it is more clearly, rather than as an opportunity to blunt the attacker's cyberweapons; this improves the fidelity of the signal.[29] Third, and

---

[27] We are not using *active defense* to apply to a capability similar to *immune surveillance* (good code that sits on routers and looks for bad code with the intent of destroying or neutralizing it). For an article that appears to assume the efficacy and centrality of active defense as a network protection mechanism, see Eric Talbot Jensen, "Computer Attacks on Critical National Infrastructure: A Use of Force Invoking the Right of Self-Defense," *Stanford Journal of International Law*, Vol. 38, 2002, pp. 207–240.

[28] Since the attacking computer is interested only in seeing return messages from the target system, it should not be particularly difficult to generate a filter that returns only screen information and thereby eliminates everything else that could harm the operations of the attacker's own computer.

[29] Chapter Four, which examines a wide variety of motives for the initial cyberattack, does list some reasons (other than deterrence or coercion) that certain types of cyberattacks may benefit the attacker directly. Except for whatever ancillary benefits may accrue from dis-

most important, if it is not possible to disarm the cyberattacker, there is little point to rushing into retaliation. More important than speed is the ability to convince the attacker not to try again. Ironically for a medium that supposedly conducts its business at warp speed, the urgency of retaliation is governed by the capacity of the human mind to be convinced, not the need to disable the attacking computer before it strikes again.

Will Third Parties Join the Fight?

To deter is to signal a potential attacker that certain acts will have undesirable consequences. As previously discussed, problems in attribution and BDA can interfere with the signal. There is another form of interference: Attacks and counterattacks may also come from third parties, thereby confusing everyone.

This problem emerges if attacks and counterattacks are visible to the hacker community. At a minimum, such an exchange would legitimize hacking to a community that is otherwise constantly lectured on how immoral and immature such activities are. After all, if states that adhere to "rule of law" do it, the only difference between legitimate and illegitimate hacking is official imprimatur. Hackers are not particularly impressed by imprimatur.

An exchange of cyberattacks between states may also excite the general interest of superpatriot hackers or those who like a dog pile— particularly if the victim of the attack or the victim of retaliation, or both, are unpopular in certain circles. The very nature of the attacks is likely to reveal the victim's general vulnerabilities (X is not impregnable) and perhaps even specific vulnerabilities (this is how to get into X). They put certain assets "in play" in the same sense that a takeover bid for a corporation makes it a feasible target for others. Both attacker and retaliator may have to face the possibility that third-party hackers

---

abling the attacker's capability to exert physical force, the explanations do not apply to the retaliator.

may continue to plague the target even after the original attacker has pulled back.

Outside participation matters because hacking is one of the activities in which third parties can play in the same league as states.[30] Software, after all, comes from the commercial world; it is broken by individual hackers and repaired by other individual hackers. It is not unknown for single individuals to break copyright locks that corporations put into the market.[31] States may have a larger panoply of attack methods than individuals do, but that is of little help in determining whether a state or an individual carried out a single particular attack.

The emergence of third-party hackers could further complicate attribution and make it difficult to understand the relationships among attack, retaliation, and counterretaliation. The prospect that attacks may continue after the attacker and the target have found out how to live with one another will complicate efforts to restore status quo conditions or even promise as much as a condition to cease hostilities.[32]

All this weakens an implied promise of deterrence: If you stop, we stop. With the existence of third-party hackers, the "we" loses its strength. What attackers want to hear—if you stop, it stops—may not be something the retaliator can promise. Fortunately, third-party attackers may strengthen an implied threat of deterrence: Do not even start because who knows where it will lead.

---

[30] By way of analogy, if cyberwar is like toxicity and if it takes a certain mass to cross a threshold, state hackers can carry out attacks that unorganized freelancers cannot. If it is like cancer, in which one well-placed mutation (however unlikely) suffices, the efficacy of a thousand-hacker state may be no better than that of a thousand networked hackers going at it full-time. To the extent that finding vulnerabilities is critical, the cancer model fits better. To the extent that the use of sophisticated tools to exploit vulnerabilities is what distinguishes success from failure, the toxicity model fits better.

[31] John Leyden, in "Blu-Ray DRM Defeated," *The Register*, January 23, 2007, reported that "[t]he copy protection technology used by Blu-Ray discs has been cracked by the same hacker who broke the DRM technology of rival HD DVD discs last month." See also Peter Svensson, "Teen 'Unlocks' iPhone with Soldering Iron," *Mail & Guardian online*, August 26, 2007.

[32] According to a former NSC staffer, third-party attacks are particularly unpredictable. Because groups outside of government can launch them, such attacks can escalate crises even as governments are trying to diffuse them (Gorman, 2008).

Does Retaliation Send the Right Message to Our Own Side?

Some potential cybertargets are government systems and some are private. The latter set includes almost all U.S. energy, communications, and financial infrastructures. Severe attacks on them are likely to get the public's attention. With a few exceptions (discussed later), government systems are not so essential to day-to-day life.

The defense of private systems is largely in private hands. Although the government can play a key indirect role in protecting such systems (e.g., through the development of policies, standards, and law enforcement), it can do little directly. The government has no privileged insight into specific vulnerabilities of private systems, and there is little evidence that private system owners are interested in telling it.

Ironically, a government deterrence policy may weaken rather than strengthen the private sector's incentive to protect its own systems if that policy alters who is responsible for third-party damage. If the power industry, for instance, fails to protect its supervisory control and data acquisition system, and it then gets hacked into and shut down, the cost to its users (i.e., blackouts) far outweighs the lost revenue to the power company. The threat that angry customers could sue the company and recover damages (or that regulators will get angry) has to be uppermost in the minds of the power company's security managers. The same holds in general for public or at least publicly accessible infrastructures.[33]

---

[33] Software makers have strenuously and successfully opposed mandatory indemnification for poorly performing products; see Todd Bishop, "Should Microsoft Be Liable for Bugs?" *Seattle Post-Intelligencer*, September 12, 2003; Michael A. Cusumano, "Who Is Liable for Bugs and Security Flaws in Software? *Communications of the ACM*, Vol. 47, No. 3, March 2004; Ira Sager and Jay Greene, "Commentary: The Best Way to Make Software Secure: Liability," *BusinessWeek*, March 18, 2002; Bruce Schneier, "Information Security: How Liable Should Vendors Be?" *Computerworld*, October 28, 2004; and Chris Gonsalves, "Security Quandary: Who's Liable?" *eWeek*, February 25, 2002. At least utilities, one can argue, have legal obligations to provide public services and can thus be more easily held to account should they fail. Furthermore, while provably correct software is nigh impossible to write, infrastructure owners can use wide array of redundant methods to minimize the risk of hacker-induced failure.

Any policy that stipulates or even hints that a cyberattack is an act of war (or even terrorism) tends to immunize infrastructure owners against such risk. A cyberattack would be considered on a par with other acts (e.g., abnormal weather) that are beyond the power of the infrastructure owner to abate. Infrastructure providers could, in effect, declare force majeure and thereby evade their obligation to provide continuous service. Similarly, persuading the public that such attacks are beyond what infrastructure owners can protect themselves against would reduce political pressure on them to keep their systems clean. Indemnification, in turn, reduces their incentives to protect their own systems.

A deterrence policy, as such, creates a moral hazard that could induce owners to postpone the vigorous search for vulnerabilities in their own systems.

Do We Have a Threshold for Response?

How bad must an attack be to justify retaliation? The defender (perhaps backed by a global consensus) can chose to retaliate against *any* intrusion into its systems that leaves systems less capable. Alternatively, it can define a threshold of damage beyond which a response would be called for.

Choosing a zero-tolerance policy is asking for trouble. If CNE is (unwisely) included, the potential for a *casus belli* will always exist, and the difference between retaliating and not retaliating will have much more to do with accidents of discovery than attack activity. If implants are (somewhat unwisely) included,[34] the crossover point will likewise be breached continually. Even if every state is fastidious about not crossing the line, can a fastidious target state afford to investigate every bot it finds to determine whether those who planted it work for some potentially hostile state? Thus, at minimum, the class of events

---

[34] If the criterion for the attack is that the use of the system is hindered, an implant, in and of itself—which may be dormant or may be used only for CNE or to make a computer into a bot—is not tantamount to an attack as the term is used here.

labeled as attacks should include only those known to involve disruption or corruption; even then, DDOS attacks may merit partial exception. To repeat an earlier point, if retaliation is more likely to follow the occasional discovery than the constant activity, the supposed attacker cannot help but ask, "Why me, why now?" and perhaps draw the wrong lessons from retaliation. The proportionality issue also weighs against a zero-tolerance policy. Minor attacks leave minor damage, generally too small to merit the attention of what would be a U.S. small claims court. Cranking up the machinery of retaliation for something so small, unless it is repeated in very large quantities, would exceed the actual damage by several orders of magnitude. Any retaliation large enough to be noticed on its own—unless announced as such—would have to be fairly large or very precisely targeted and would therefore be viewed as much more serious than the original infraction. If the target did not believe retaliation of such magnitude was deserved and therefore also responded disproportionately, escalation would loom. Strict adherence to a no-threshold policy of response also implies a no-threshold policy of investigation of cyberattacks, one that is untenable and, in any case, unaffordable.

True, a zero-threshold policy has one big advantage: A state could demonstrate its will to retaliate for large attacks (that have not happened yet) by retaliating, even if in lesser measure, for small attacks. But the smaller the attack, the smaller the signature. Although it is possible to argue that very large attacks can come only from states, no such relationship covers small attacks. It is unlikely that the attacker will confess to a small charge; no state has yet to own up to conducting CNE, much less a small cyberattack. Ditto for BDA; if retaliation for small attacks is correspondingly small but undertaken just to prove a point, the attacker (as the target of retaliation) may have a problem determining that it was, in fact, retaliated against. The smaller the disruption, the more likely it is to have looked like an accident. Even if the attacker received some specific indication that the attack was an act of retaliation, such signals may not necessarily reach the attacker's public (or the third parties the retaliator wishes to impress). Small attacks may, anyway, appear to lend themselves to prosecution rather than retaliation because they look like the acts of a single person or a small group.

Thus *any* retaliation could seem provocative. Meanwhile, the attacker may have learned a cheap lesson from low-level retaliation—not about the foolishness of cyberattacks but about the importance of covering one's tracks. It may also gain insight about what kind of attacks are likely to be caught or, if it is lucky, something about the forensic methods the target uses. Still, in a zero-threshold posture, this may inhibit the aggressor's contemplation of a cat's-paw gambit: The target did not retaliate against *this*, so I will try *that* next. If adversary believes that it can carefully calibrate increasing levels of attack—extremely hard to do in practice—it may hope to replicate what happens to frogs put in slowly boiling water: No gradient of added pain is sharp enough to make them jump out.

Unfortunately, selecting and monitoring activity against any one threshold is no picnic either, even if the state allows itself and the suspected attacker wiggle room. Loss of life might be one threshold; in terms of clarity, death has the advantage of being unambiguous. The U.S. strike against Libya in 1986 was justified as retaliation against an allegedly Libyan-sponsored bombing in Berlin that killed two Americans. Few in the United States thought that this was an arbitrary flash point.[35] Yet cyberattacks can kill people only as a secondary, rather than primary, consequence, and 20-plus years of cyber mischief have yet to claim their first clear casualty. Chances are that the first casualty from a cyberattack (unless it takes place in the context of war) is likely to come because some accident was made more likely or because some warning and control system was knocked offline. Given the indirect chain of events cited here, justifying retaliation based on such an event would hardly be simple.

Economic criteria—e.g., retaliation will follow if the attack cost more than $1 million—are tractable, and offer the promise of some reasonable proportionality, but are hard to define. How does one put

---

[35] Mark Whitaker and John Wolcott, in "Getting Rid of Kaddafi," *Newsweek,* April 28, 1986, reported that "Early polls show overwhelmingly popular enthusiasm for the president's decision to punish Kaddafi and, publicly, administration officials are confident that support will hold up." It is quite another question whether the strike on Libya was proportionate or even wise, given that the Lockerbie incident, which killed more than 200, was almost certainly an act of counterretaliation.

a price on lost secrets, lost privacy, or lost trust (admittedly, these are consequences of CNE more than they are of cyberattack)? An attacker could easily cross the threshold by accident—although, in legal terms, this is not much of a mitigating circumstance (someone who takes a deliberate but random shot into a sparse crowd and, contrary to odds, kills someone can be indicted for murder). The potential retaliator would have a double burden: not only establishing causality between an attack and the subsequent damage but also, unless the threshold was low or the damage clearly high, making a convincing case that the damage exceeded the threshold. There would also have to be some consensus about how to measure the cost of monitoring the attacked system for nonobvious damage and putting in additional safeguards to prevent the next such attack. Such expenses are not inherent in the attack but are decided on afterward by the target.

The economic threshold problem can be mitigated by requiring, say, a ten-to-one ratio between measured damage and the threshold, leaving room for error. This only works if the number of attacks whose damage exceeds the nominal threshold (i.e., $1 million worth of damage from all related attacks) but not the actionable threshold ($10 million) is small, so that retaliation against the big attack does not seem arbitrary when other attacks that could have crossed the announced threshold are ignored.

The threshold question pertains even if states respond to cyberattacks with admonishment rather than punishment. A norms-based threshold would require defining, or helping the community of nations define, what bad behavior is (e.g., hacking is not simply "boys will be boys" or "spies will be spies"). But with cyberretaliation, the stakes in getting it right are higher, and the arguments about proportionality may surface only after it is too late to take things back.

Can We Avoid Escalation?

Nuclear deterrence strategists did not worry about escalation beyond the nuclear level.[36] If the attacker had used nuclear weapons, it was hard to argue that retaliation would induce them to do something much worse than what they had already proved they were willing to do.

Cyberdeterrence strategists *do* have to worry about such issues. The attacker may respond to retaliation by escalating into the violent or even nuclear realm. Indeed, for a while, it was Russia's declared policy to react to a strategic cyberattack with the choice of any strategic weapon in its arsenal.[37] Attackers are likely to escalate if they (1) do not believe cyberretaliation is merited, (2) face internal pressures to respond in an obviously painful way, or (3) believe they will lose in a cyber tit-for-tat but can counter in domains where they enjoy superiority.

Attackers are also likely to escalate if the retaliation crosses a threshold in their own perception—even if the retaliation is in kind and appears proportionate to the retaliator. Accidental and inadvertent escalation exists in the real world.[38] In cyberspace, where the ultimate effects of the attack are so uncertain and the ground rules are practically nonexistent, the risks are even higher.

Attackers could threaten physical counterretaliation in hopes of reducing the credibility of the target to that of a bluff. Those who would forestall a cyberattack by threatening retaliation in kind may lose to an attacker who counterthreatens escalatory counterretaliation.

---

[36] Although they did have to worry about escalation within the nuclear level. For example, would a state that had dropped a nuclear weapon on Europe next attack the United States if the United States responded by attacking Soviet interests? Would a nuclear response to a nonfatal nuclear detonation (e.g., to create a high-altitude electromagnetic pulse for frying electronics) lead to a fatal nuclear counterresponse? In 1965, Herman Kahn identified 44 steps on the escalation ladder, of which 29 were at the nuclear level (Herman Kahn, *On Escalation, Scenarios and Metaphors*, New York: Praeger, 1965).

[37] Blank, 2008; Stephen Blank, "Can Information Warfare Be Deterred?" *Defense Analysis*, Vol. 17, No. 2, 2001, pp. 121–138; and Matthew Campbell, "'Logic Bomb' Arms Race Panics Russians," *The Sunday Times*, November 29, 1998, as taken from Blank, 2001.

[38] See Forrest E. Morgan, Karl P. Mueller, Evan S. Medeiros, Kevin L. Pollpeter, and Roger Cliff, *Dangerous Thresholds: Managing Escalation in the 21st Century*, Santa Monica, Calif.: RAND Corporation, MG-614-AF, 2008.

Incidentally, any state that carries out a seriously damaging cyberattack on a nuclear-armed state necessarily runs the risk—small, perhaps, but not zero—that nuclear war may result from its actions. If it decides to attack regardless, it may be because it believes that the benefits from attacking (or the costs from *not* attacking) are sufficiently large. If the attacker believes that the benefits merit running the risk of nuclear war, how much would it be daunted by the additional (albeit more plausible) risks of cyberretaliation?[39]

What If the Attacker Has Little Worth Hitting?

Perfectly symmetric warfare does not exist, particularly when the United States is involved. Yet cyberwarfare may be more asymmetric than most.[40] The U.S. economy and society are heavily networked; so is its military. The attacker, by contrast, may have no targets of consequence, either because it is not particularly digitized, because its digital assets are not networked to the outside world, or because such assets are not terribly important to its government. The DDOS attacks that knocked out servers in the well-wired nation of Estonia (or "E-stonia" as some of its countrymen like to boast) in May 2007 were greeted with shock. Those against Georgia (August 2008) were greeted with some dismay. Finally, the January 2009 attacks against Kyrgyzstan, in central Asia, were hardly noticed at all. Conversely, when unwired states do get digital equipment, they tend to buy it from others, which makes them potentially vulnerable to supply-chain attacks, a hit-or-miss proposition that may reverberate (by ruining the vendors thought responsible for the damage).

The prospect of retaliating after a cyberattack from a target with nothing important to lose is not pleasant. If nothing else, this makes

---

[39] The basic argument is from RAND colleague David Frelinger. The attacker could figure it can run the risk of nuclear escalation because the odds of getting caught are low, but such low odds equally vitiate the deterrence value of cyberretaliation.

[40] Similarly, space warfare is not very useful against those without satellites. The difference is that space assets might conceivably be used to attack earth targets, while the tools of offensive cyberwar have no other use.

a declaratory policy that contemplates staying in the cyberlanes look a little foolish. The results of going ahead with such a policy may be completely successful yet altogether trivial. If the United States escalates using kinetic means, it loses whatever advantage it may garner from keeping others in cyberlanes and must explain why it was the first to use violence (unless the opening cyberattack itself created casualties).

Yet the *Will* to Retaliate Is More Credible for Cyberspace

A key paradox of nuclear deterrence arises from the question of whether someone who threatened retaliation would, in fact, carry it out when the time came.[41] The very concept of deterrence presumed that would-be attackers were rational and that they, being rational, would conclude that whatever was to be gained by aggression would be more than overwhelmed by the nuclear retaliation that followed. The problem in such a formulation is the presumption that the victim of aggression had to be at least somewhat irrational, particularly if there was valid cause to believe the original aggression had limited scope. The gains from retaliation—such as having one's threats be taken seriously—paled before the destruction that might arise if both sides started emptying out their nuclear arsenals on one another. Yet if the aggressor believed that the victim would not act irrationally and retaliate, deterrence could fail.[42] Several strategic scholars tried to deal with the problem in different ways. Thomas Schelling argued for a deterrence that "left something to chance." Herman Kahn argued that building enough bomb shelters

---

[41] In this respect, nuclear deterrence differed sharply from conventional deterrence, even in the airpower era. Someone attacked by a bombing raid could assume war had already begun and that this war would be decided by other means, such as ground power. Thus, absent solid information that the air raid was an accident or a one-off event, the case for retaliation was strong. See Quester, 1986, esp. pp. 136–158.

[42] Some U.S. strategists argued in the late 1970s that the Soviet Union could plausibly take out the land and air legs of the U.S. triad in a first strike, spare U.S. cities, and leave the United States to choose between strategic inferiority or losing its cities. Conversely, John Mueller argued that the prospect of fighting a large-scale conventional war was what deterred the major powers in the Cold War. See John Mueller, *Retreat from Doomsday: The Obsolescence of Major War*, New York: Basic Books, 1989.

could make a U.S. threat to fight a nuclear war more credible. Patrick Morgan argued that "rationality" may be the wrong standard and that thinking in terms of a "sensible," rather than a rational, deterrence policy might avoid some of the conundrum.[43] Nevertheless, the question of "will" was central to strategic thought, and there was also considerable debate on how a state would express its will to retaliate before actually having to do so.

The question of will is not absent in cyberspace, but it is much less of an issue, especially if the odds are sufficiently high that war in cyberspace can be decoupled from more-violent forms of war. The key difference between nuclear deterrence and cyberdeterrence is that a full-fledged cyberattack may be burdensome, expensive, and highly unpleasant but also survivable. When the registers clear, systems will be reestablished, and deterrence, if it makes sense at all, would have to be reestablished for the next time. Thus, it could very well be rational to retaliate because it would help later and would boost the credibility of other deterrents, as Appendix B suggests—especially if the retaliator had declared its intentions beforehand.

Conversely, rejecting retaliation, even after declaring the intention to use it, does not bespeak irrationality. For example, a specific instance of refusing to retaliate might be due to the realization that this particular attack succeeded because of some oversight on the part of the target's sysadmins, that they would correct the mistake, that no such attack would henceforth take place, and that therefore this one incident could be ignored. The alternative would be committing to a confrontation with the attacker that might lead to further damage all around—not just in cyberspace.

Thus, while attackers may doubt the target state's willingness to retaliate, such doubts and how to resolve them in the opponent's mind do not play the central role in cyberdeterrence that they do in nuclear deterrence. Of greater importance may be whether or not the target state has, in fact, retaliated against an attack. As Chapter Five describes in more detail, a failure to retaliate could be a failure of will, a failure

---

[43] See Patrick Morgan, *Deterrence, a Conceptual Analysis*, Beverly Hills, Calif.: Sage Library of Social Research 40, 1977, esp. pp. 103–126.

to detect the attack (e.g., a corruption attack), a failure to attribute the attack, or a failure of the response to register.

A Good Defense Adds Further Credibility

Although a successful cyberdeterrence posture can be justified by the money that one can save on defense, cyberdefense adds credibility to a cyberdeterrence posture.

The first effect is straightforward: The better one's defenses, the less likely it is that an attack will succeed and so the less often a cyberdeterrence policy will be tested. The longer such a policy goes untested, the more credibility it acquires, if only through precedent.[44]

Second, a good defense adds credibility to the threat to retaliate, much in the way Herman Kahn argued that having bomb shelters made nuclear deterrence more credible. Likewise, demonstrating the ability to absorb counterretaliation without flinching increases the likelihood in the attacker's mind of being retaliated against because the costs of sparking at full-fledged cyberwar would fall disproportionately on the other side. Unfortunately for the analogy, such credibility tends to be associated with what Kahn labeled Type III deterrence: the ability to get one's way on nonnuclear matters (e.g., a conventional attack in Europe) by threatening nuclear action. Finding an analogy in cyberspace requires identifying issues *below* the cyberwar level, where the threat of escalation to cyberwar could decide the issue in one's favor—a prospect that may be defeated by the many uncertainties and ambiguities of cyberwar.

Third, good defenses have a way of filtering out third-party attacks, *if* third parties are incapable of rising to the level of sophistication of state attackers. This means that one argument against retaliation—putting one's infrastructure at risk from emerging hackers—loses much of its force.

---

[44] This relationship, however, would be weaker if the policy were to retaliate after failed attacks. Because of their premature termination, these leave even less forensic evidence than do successful attacks, so that attribution is likely to be harder for failures.

Fourth, to the extent that good defenses filter out third-party attacks, they facilitate attribution by elimination. This should not be overstated, since the differences between state and nonstate attackers may be subtle, and the problem of distinguishing between attacks from two or more cybercompetent states (e.g., was it Russia or China?) remains.

CHAPTER FOUR

Why the Purpose of the Original Cyberattack Matters

The first question a potential retaliator must ask is what the attacker was trying to achieve with its (presumably unprovoked) attack. Answers may indicate how legitimate and wise retaliation is and whether other strategies need to be pursued. From a strategic perspective, having a good idea of why a state carried out a cyberattack offers some insight into its decisionmaking calculus. Understanding what it stands to gain or lose through an attack helps immensely in figuring out what kind and level of retaliation—if any—can tip the attacker's thinking away from initiating or continuing cyberattacks. Its motivation will also color how the attacker perceives retaliation. For instance, the less legitimate an act of retaliation appears to the attacker, the more likely it is to counterretaliate. Thus, motivation informs the decision to retaliate. As this chapter demonstrates, there are many, very different motives for what is (or looks like) an unprovoked cyberattack.

This chapter examines four classes of motive: error, coercion, force, and other.[1] For each class, the sections below offer a brief description and a discussion of some choices a potential retaliator may wish to consider.

---

[1] Here, *error* includes both error on the part of the attacker and on the part of the retaliator.

Error

The attribution may be correct, but the presumption that the attack was a deliberate first strike by a state may not be.[2] The attacker's command authority may not realize it has, in fact, been attacked. The attack may have been an accident. Or the attacker may view the event as an act of retaliation, even if it is not. The following paragraphs examining these cases.

Oops

The attack could actually be an accident or, less defensibly, may have been a minor flick that accidentally crossed a threshold.[3] The attacker may have attempted to steal information or test the target's systems' reaction to a partial cyberattack but its efforts ended up wreaking damage.[4] Or there might have been an unexpected interaction between

---

[2] Erroneous retaliation presents some obvious difficulties. At what point, if any, does the retaliator, for instance, admit it erred? After the United States shot down an Iranian Airbus in 1988, then–Vice President George H. W. Bush proclaimed, in effect, that superpowers do not apologize. But the United States did pay compensation afterward, and the mistake was obvious. Neither apology nor compensation followed revelations that the United States exaggerated the North Vietnamese role in the Gulf of Tonkin incident (which lead to retaliatory air strikes) or that Iraq did not, as advertised, actually have weapons of mass destruction (which justified an invasion).

[3] Less plausibly, an initial attack might be the result of a target trying to remove or route around an implant that had been booby-trapped to prevent just such actions. Such an attack would be a highly aggressive complement to espionage, a motive that needs no further elaboration. It is unclear why any rational attacker would do this. The whole point is to keep the implant hidden, something at odds with the noisy revelation that would occur if an implant were jostled. If the booby trap were announced in advance (otherwise it would be one of the chairman's surprises, as in the movie, *Dr. Strangelove*) doing so would be judged quite hostile.

[4] Had this report of a February 2008 blackout had been correct, it would have provided an example:

> A Chinese PLA hacker attempting to map Florida Power & Light's computer infrastructure apparently made a mistake. "The hacker was probably supposed to be mapping the system for his bosses and just got carried away and had a 'what happens if I pull on this' moment." The hacker triggered a cascade effect, shutting down large portions of the Florida power grid, the security expert said. (Shane Harris, "China's Cyber Militia," *National Journal Magazine,* May 31, 2008.)

the breaking-and-entering activities and the defense mechanisms of the target system. Here, both sides face dilemmas: The attacker cannot claim that its efforts have gone awry without admitting that it was trespassing; whatever retaliation it avoids by proving the first may be more than made up by admitting the second. In the absence of confession, the potential retaliator will have no good way of knowing whether the attacker actually thought the attack was an accident. If the attacker confesses to having erred but still suffers retaliation, it may feel that the retaliator went out of bounds and may then counterretaliate.

No, You Started It

Attackers who believe themselves to be righteous retaliators may be incorrect in their self-assessment (e.g., their attribution was bad), correct (e.g., the target's leadership was unaware that it had attacked), or oversensitive.[5] In the last case, the oversensitive attacker may believe that the target crossed some line, perhaps exceeding a very low threshold or carrying out an act not traditionally held to be a *casus belli* (e.g., the target's support for an economic embargo against the attacker). In all three cases, if the target then retaliates, the attacker may well view retaliation as unjustified, a continuation or even escalation of the prior attack—and perhaps even more illegitimate than the previous (perceived) attack. Indeed, counterretaliation may confirm to the attacker that it was not in error when it "retaliated." Whether or not the attacker would counterretaliate (to the attacker, this would be the fourth round) may rest on nothing more reassuring than whether its fears of descending into a cyberwar exceed its umbrage at having been unfairly attacked and then retaliated against for responding.

But the report was incorrect; a transmission system element fault was the cause (FRCC Event Analysis Team, "FRCC System Disturbance and Underfrequency Load Shedding Event Report February 26th, 2008 at 1:09 pm," final report, Tampa: Florida Reliability Coordinating Council, Inc., October 30, 2008).

[5] We assume that the target's leadership is unaware that it attacked first. Otherwise, it would recognize the attacker's move as retaliation, which would need no further explanation.

Rogue Operators

An attack could come from within state organs but not from the state. The instruments of cyberattack are neither so enormous as to require national command authority nor so obvious as to subject their use to state veto. Even the fact that intelligence for the attack was collected under official auspices does not prove the attack was authorized: Intelligence preparation of the battlefield may have taken place as a contingency for cybercombat at a later time or for combat using other means. Intelligence could have seeped to the actual attackers, who could be rogue bureaucrats, organized criminal enterprises, coteries of well-connected hackers, or superpatriots (e.g., the sort that defaced Web sites in China and the United States in spring 2001).[6]

The Command-and-Control Problem

Accidents and rogue attacks both illustrate the command-and-control problem in cyberspace. In the physical world, it is easier to distinguish accident from attack,[7] and overseeing the use of weapons is relatively straightforward. Not so in cyberspace. Comparable oversight requires supervisors to look over each operator's shoulder and figure out what is going on—when one keystroke looks much like another.[8] The effects of such actions on target systems are even harder to monitor in real time. Because cyberspace remains arcane to conventional warfighters, much less to their political leaders, much of what operators may report can get lost in translation. All this multiplies the scope available to rogue operators or even cowboys. Leaders may not know with any confidence what mischief has been perpetuated in their name against other people's systems. The cyberwarriors may not know themselves; the leaders

---

[6] See Michael Reilly, "How Long Before All-Out Cyberwar?" *New Scientist*, No. 2644, February 20, 2008, pp. 24–25, and Rose Tang, "China Warns of Massive Hack Attacks," *CNN.com*, May 3, 2001.

[7] But not trivial. In World War II, the issue of whether civilians were fair game for bombers was accidentally resolved when German bombers dropped their loads on residential neighborhoods when they had intended to limit themselves to industrial targets.

[8] Somewhat less supervision is required if operators do no more than run scripts and monitor the effects of the scripts. But it is far easier to protect a system against what are called "script kiddies" than against wily hackers.

may not understand them; and, even if the leaders do understand, may not necessarily trust the cyberwarriors when they say "oops."

It should not be hard to argue in favor of diplomacy and explicitness as an alternative to assumptions and retaliation in such cases.

Coercion

The biggest difference between coercion in real space and cyberspace may be one of credibility.[9] If a bully rolls thousands of tanks up to your border and announces its desire that you accommodate its interests, you may well consent without forcing him to demonstrate that the tanks are as capable as they look. The same credibility calculus does not work in cyberspace. You may be entirely unsure of what a cyberattack may do to your economy and society because you are unsure of how capable the bully is and how vulnerable you are. Indeed, because its threats presume your vulnerabilities, you may not believe that the bully knows more about the vulnerabilities in your state's systems than your people do (and if they knew about them, they would fix them). The fact of the threat may similarly suggest that the bully knows something that you should know; this threat may be a wake-up call to install all the current patches, make another scrub for vulnerabilities and aberrant system behavior, and reexamine earlier decisions to open systems up to the outside. With all that uncertainty, the bully may have little choice but to show what it can do if it wanted to have much chance of coercing you at all. Hence, the attacker needs to strike to demonstrate some coercive capability.

Will one attack do? Maybe not. Most attacks exploit vulnerabilities.[10] If, following the coercive attack, the specific vulnerability (or class of vulnerabilities) can be identified and fixed in every system that

---

[9] Although it is natural to assume that people resist coercion simply because they do not like to be perceived as capable of being pressured, coercion may be considered just another set of created incentives. The threat, "do X or you will be bitten by the dog," is equivalent to "if you do X, I'll intercede with the dog and persuade it not to bite you." The latter sounds slightly friendlier.

[10] Flooding and DDOS attacks, as usual, aside.

mattered, the party to be coerced might feel that it has regained its invulnerability and hence resistance to coercion. It would thus take a second attack—but then, the same cycle of discovery, patch, and resistance might require a third, and so on (note the similarity to the discussion in Chapter Three about retaliation). On the other hand, one attack might suffice because of psychological factors. The first three of the four September 11th airline hijackings succeeded because of a security "vulnerability": The hijackers exploited passengers' presumption that they would survive the hijackings if they did not resist. Passenger resistance to the fourth hijacking demonstrated that such a "vulnerability" closed that day. Nevertheless, the attention paid to airline security then rose sharply, even though the susceptibility to hijacking was lower the week after 9/11 than the week before.

To coerce, an attacker must signal that a specific set of cyberattacks was meant to coerce. *What kind* of coercion is another question. Although the target may learn several lessons from the attack—about vulnerabilities it was insufficiently aware of or that the attacker is a bully—it may learn nothing about what the attacker's vital interests are. Hence, the attacker has to assume that the target already understands which of the attacker's interests are vital and that contravening such interests comes with risks.

The attacker's corollary challenge, one rarely present with physical attacks, is to associate itself with the cyberattack. Calling the shot beforehand will leave no doubt that it was the one,[11] but if the cyberattack fails, the attacker's credibility will flounder. Taking credit afterward may require the attacker to do something specific to prove to the target that it, not others, really *was* responsible. Anything less than an explicit link may leave questions. While a cyberattack in the midst of an ongoing crisis might seem to suffice, it was such logic that led Estonians to prematurely finger the Russian state in 2007. So, attribution may be less than obvious in future cases. Fortunately for coercion,

---

[11] Normally, warning the defender even minutes ahead of time can permit certain defenses to be taken (e.g., pulling systems offline). Yet a little imagination suffices to suggest many ways of establishing credibility after the fact, such as a letter mailed beforehand but received afterward, a "Kilroy was here" leave-behind in the target system, or revelation of knowledge about the target system that only penetration would provide.

perfect certainty is not required; less-than-obvious linkages may still inhibit the target's desire to trespass on the attacker's interests.

Attacks used to impress third parties require visible effects.[12] An attack that turns off the lights in the target's city would qualify. One that scrambles currency transaction records or the target's military logistics database may not (or at least not immediately), even if more damaging. Again, this trade-off is muted with physical attacks.

Because attribution is likely to be better and because legitimacy is less problematic when the motive is coercion, retaliation can proceed on a firmer footing. This does not mean that the attacker will not scream anyway. Its supporters may believe that anything it does is justified and that anything done to it is not or that blowing enough smoke will convince people there is a fire somewhere. If attribution is not backed up by good evidence, the attacker may wish to protest its innocence, truth to the contrary. Yet an attacker should expect that retaliation is possible but not certain—especially if the target is intimidated—which was the goal of the attack in the first place. Presumably, the attacker took the target's ability to retaliate into account when it pondered its coercive attack in the first place—then did it anyway. The onus is then on the target to ensure its retaliation exceeds the attacker's expectations (unless the attacker expected nothing coming back and thus any retaliation would exceed expectations).

The narrative of coercion (via attack) and resistance (via retaliation) can be misread in both directions.

Consider, first, the use of coercion to gain advantage in a crisis that involves other issues (e.g., the Chinese making a play for Taiwan) by signaling to outsiders that intervention would be costly. Facing attacks on civilian systems, the target's leadership may conclude that the real issue is not proximate (e.g., Taiwan) but strategic: its ability to withstand coercion. The proximate issue then becomes a proxy test of strength for the strategic issue (e.g., resisting Chinese aggression). By elevating its proximate conflict's importance in the target's mind, the attack elevates the cost of the target's *not* intervening in the proxi-

---

[12] So, the "Kilroy" methods that identify the attacker to the target would not suffice.

mate issue (looking as if it is yielding to a direct challenge) and thereby makes intervention more likely.

Consider, next, the possibility that what looked like coercive attacks were instead meant to *elicit* retaliation. A state may be so motivated if it thinks war inevitable but prefers to be viewed as the victim rather than the aggressor. It thinks that important opinion makers will react more to retaliation, especially if disproportionate, than to the original provocation. This logic may have motivated Bismarck to manipulate Napoleon III into foolishly declaring war against Prussia in 1870.[13] Nonstate actors may find such logic even more appealing. Although the September 11th attack may have looked like a coercive act meant to frighten the United States into ceasing its interventions in the Islamic world, the opposite occurred. Perhaps Osama bin Laden, who saw the Soviet Union mired in Afghanistan and collapsing thereafter, hoped for a similar fate for the United States; as a bonus, conflict between the United States and elements of the Islamic world would polarize the Islamic world and rally fundamentalists around al Qaeda.[14]

Force

Cyberattacks on a target's military and related systems are usually meant to weaken the target's ability to respond to crisis. A large, successful attack may retard the target's ability to wage war; if the target's military deployment can be delayed long enough (e.g., after everything has been decided and after the aggressor's forces have dug in for defense), the target's military intervention may be deemed pointless.[15] This may well be *the* key cyber risk: An attacker convinces itself that an otherwise infeasible military attack can be made feasible by a bolt from the blue. It (e.g., China) starts a war (e.g., to take Taiwan) on

---

[13] Michael Howard, *The Franco-Prussian War*, London: Routledge, 1961.

[14] See, for instance, Rohan Gunaratna, *Inside Al Qaeda's Global Network of Terror*, New York: Columbia University Press, 2002.

[15] See for instance, James C. Mulvenon, "The PLA and Information Warfare," in James C. Mulvenon and Richard H. Yang, eds., *The People's Liberation Army in the Information Age*, Santa Monica, Calif.: RAND Corporation, CF-145-CAPP/AF, 1998, pp. 175–186.

the assumption that, with networks down, the target (e.g., the United States) cannot intervene until too late unless it wishes to ignite a general war. Even if the attacker is wrong, everyone suffers.

The key question, when the dust clears, is when the shooting war is coming.[16] If the cyberattack is today, for instance, but the expected engagement is five years hence, the useful effects will have long been reversed by the time of the engagement. Disruptive attacks help sysadmins identify the vulnerabilities that were exploited. Corruption attacks may take longer to find, but every day that passes creates the risk that they, too, will be discovered and that the vulnerabilities that permitted the attacks can also be fixed. Even if the vulnerabilities are not discovered, normal software turnover may also eliminate them. Because vulnerabilities, once fixed, are harder to exploit, the benefit for the attacker is likely to be negative by the time war breaks out months or years hence.

Thus, military operations following such an attack would have to start in the hours or, at most, days before the target can discover and repair the attack and thereby reemerge little worse off or even stronger. Indeed, hours (or minutes) make more sense than days because a cyberattack, if recognized as such and if attributed correctly, would toss away the advantages of a surprise physical attack to achieve whatever advantages surprise bestows on a cyberattack. Conversely, a failed cyberattack may convince the attacker that it would be foolish to go to war if the target's military information infrastructure is still intact. So, no such attack occurs.

In such circumstances involving the United States, retaliation would be no higher than the fourth issue on the President's plate. First would be determining whether war were, in fact, imminent—how soon

---

[16] This does not apply to disruptive cyberattacks against continuous surveillance systems for the purpose of creating coverage holes, during which there are opportunities for mischief (e.g., transporting nuclear materials). India's nuclear program, for instance, supposedly evaded U.S. oversight because work that had to take place in the open was scheduled for hours when surveillance satellites were not overhead. That noted, intelligence systems are far more likely to be self-contained than military systems, and the difficulty of disrupting them through cyberattacks is, or at least should be, virtually zero. If attacks are successful, the target has important internal matters to worry about; retaliation is secondary.

and by what means. Second would be recovering the posture of the affected military units. If an attack were deemed inevitable, the highest priority would be to restore as much capability as quickly as possible to be ready against the hour or day of the attack (e.g., favoring patch-and-recover over deep-cleaning). Third would be conveying readiness if the fact and timing of the attack appeared contingent on how much damage the attacker thought the cyberattack had caused. The target should try to convince the attacker that damage had been minimal and was being repaired quickly. It might also wish to give the attacker the sense that it has undetected backup capabilities (but without specifying them). Success at showing a good face might convince the attacker that whatever gains it hoped to see from the cyberattack were modest and short lived. Indeed, before the expression of such a threat, the target might want to conduct and "leak" the results of exercises that suggest its military can "fight through" system failure.

Retaliation, in this context, requires reflection. Although the target should *understate* how badly it has been hurt, retaliation suggests that real pain was inflicted. The greater the retaliation, the greater the pain. The wise course is to delay retaliation until the cyberattacker has decided whether or not to wage physical war. Should war follow, both sides would be trying to hurt each other in serious ways, and cyber-retaliation might appear quite weak in comparison. If both sides were at war, everything might be locked down anyway and thus be less open to attack. Should war not follow an initial cyberattack, cyberretaliation, especially if painful, might bring on the physical attack. After all, the attacker has presumably revved up its military in preparation for imminent battle in the first place and thus would enjoy an initial military advantage—and knows it. Retaliation might also persuade the attacker not to prepare for an attack next time without carrying it through. Otherwise, having pulled back from a physical attack, the attacker would find the battle joined anyway, in cyberspace, with escalation to violence likely to take place once the target's military is ready. Alternatively, such retaliation might persuade the attacker to prepare for a physical attack using less-detectable cyberattacks the next time.

Retaliation is somewhat more attractive if the cyberattacker is not the same state as the physical attacker, e.g., the latter is a proxy for the former. In this case, retaliation is more likely to be noticed once the shooting starts because the cyberattacker, not being involved in the war itself, will not be suffering from war's effects. The credible threat of retaliation might therefore persuade potential cyberattackers to think twice before putting their hackers at the service of another state's military goals. Yet many of the other cautions still apply: Retaliation signals pain and may draw the physical forces of the cyberattacker into conflict (although these physical forces may not necessarily be as ready for war as the proxy state's forces are).

Complicating this logic are attacks that *look* like they are meant to cripple another's military but are not. For instance, what if they were meant to persuade the target military that war was imminent, draw it to the ramparts for no reason,[17] and repeat the cycle often enough to exhaust the target? In contrast to physical feints, however, cyberfeints may be poor strategy. By hardening the target's systems, every attack makes a subsequent attack more difficult. The choice of targets, if not masked by noise, may also suggest what the attacker finds important to disrupt and thus hints at how the cyberattacker would fight if war turned physical. Thus, while retaliation may be called for, so that the target is not the only one being exhausted by the games, the weakness of the strategy from the attacker's point of view suggests that the target should think carefully before concluding that it is, in fact, seeing a feint.

Attacks may be launched on military systems to see how well their operators react, in preparation for some later, larger attack. Attackers would be asking many questions. Can enemy sysadmins determine what happened and why? What workarounds do they use? Will corruption be detected? If the target knows it has been so tested, would it retaliate? Conversely, attacks may well reveal a great deal about the attacker and what it knows about the target's vulnerabilities. The target, if it understands the purpose of the attack in time, may want to react

---

[17] Egypt's strategy in the October 1973 war presupposed that the Israeli army would treat its advance toward the Sinai as yet one more exercise that could be ignored safely. See Chaim Herzog, *The Arab-Israel Wars: War and Peace in the Middle East from the War of Independence Through Lebanon*, New York: Random House, 1982, pp. 233–239.

in ways that deliberately leave a particular impression in the attacker's mind. The target may want to look more prepared than it actually is, the better to dissuade a major cyberattack following confidence in the results of a minor one. Should a major attack appear inevitable, it may want to demonstrate weakness in a particular area, perhaps leading the attacker into a trap on a real battlefield (although that may be too clever by half). Retaliation, by contrast, has a tendency to reveal that the target was hurt enough or at least outraged enough to strike back and may provide the attacker better BDA than it would have gotten from direct observation alone.

Perhaps the whole point of the attack is to make the target extra wary of expanding or opening up its networks, especially to outsiders, such as allied militaries, other government agencies, or support contractors. Further wariness may result from making the attack appear to come from a trusted source. Such a strategy presumes a skewed response from the target: not that networking should not be done naively but that networking is bad. It is easy to see why such a strategy can backfire; the target may respond by keeping the networks but hardening them. As for the target, a public face that regards these attacks as pinpricks or, at worst, birth pangs, helps demonstrate the resolve to make continuing investments in networking. Retaliation communicates otherwise; thus, it, too, can backfire.

Other

Now consider some poorly considered but not necessarily impossible motives for a major cyberattack.[18] In many cases, the attacker may have

---

[18] There may also be motives for smaller cyberattacks. The attacker may be practicing preemptive defense against a computer it thinks will attack it (even if this does nothing but temporarily disable a few hundred dollars worth of hardware). Or the attacker may want to silence a target state by cutting off its communications systems (Georgia suffered that fate for a few hours during the initial period of the 2008 Russian attack; Espiner, 2008b), but there are relatively cheap workarounds if a potential target anticipates this being a problem. The attacker may want to disable foreign computers that have been converted into bots (vigilante justice, but one made somewhat less incredible if the vigilante goes after computers in its own state first).

something to gain, but such gains are often illusory, while the cost of getting caught is not. Although retaliation does make sense in such cases, as a reality check, simple exposure may suffice and may avoid some complications in the process.

For instance, the attacker may believe it can get away with an attack and thus needs no good motive (perhaps apart from wanting to train its cyberforces on a real target). This would seem to be the easiest case for retaliation. The attacker has little to gain from carrying on, and retaliation smashes the easy assumption that such attacks can be carried out with impunity. Yet even here there are complications. Legitimacy may not be a real issue to a fair observer, but the attacker may have to pretend that it was. Indeed, retaliation may persuade the attacker to invest its original attack with a more-serious purpose in retrospect and thereby counterretaliate. Even in cyberspace, it is possible to make something of nothing.

Almost as feckless is the attacker's desire to create damage for its own sake.[19] The attacker may not be at war with the target but may believe that any harm to its target constitutes a benefit for itself—a zero-sum game. Retaliation is a tricky business. The attacker may well believe the target thinks as it does, that both are locked in a zero-sum game. Thus, the target is expected to attack (retaliate) whenever the costs of doing so are greater for the original attacker than they are for the original target. In this calculus, retaliation is just one more attack. The purpose of retaliation, ironically, is to convince the attacker that the target *does not* believe it is locked in the game. If the target attacks only just after the attacker does, the attacker may conclude the two are related, voiding the assumption that the target attacks at will. Even though the attacker's net loss is the target's net gain in a zero-sum world, the target will weigh its own pain more highly than it weighs the attacker's pain. The attacker could then conclude that future forbearance on its part will persuade the target likewise. Better yet, if retaliation stops, even though its efforts are less burdensome than the pain its causes the original attacker, the latter may conclude that the

---

[19] Consider Iraq's decision to set Kuwait's oil fields afire after retreating from that country in 1991.

target (as retaliator) is not obsessed with relative power. Perhaps the attacker need not obsess either.

A less-irrational motive for cyberattack is to raise the comparative position of the attacker. A cyberattack could convince third parties, for instance, that the target institution cannot be a trustworthy partner in cyberspace. Here, the consequences of the attack must be visible. Alternatively, attacks may divert time, attention, and resources to repair the damage and prevent further attacks. Thus hobbled, the target cannot compete with the attacker or its institutions. This is a dicey move.[20] In the civil realm, at least, attacks would likely backfire if there were any good hint of where they came from. The spat could reduce the competitive positions of both attacker and retaliator vis-à-vis third parties. Since the gain to the attacker is usually likely to be less than the loss to the target, retaliation should persuade the attacker that the net rewards are negative *if caught*. Unfortunately, this does not mean that the attacker will step back in the face of retaliation, for all the reasons cited above, e.g., doing so will admit guilt and weakness.

A variant motive is to use attacks to make citizens lose faith in the target's government. Yet (1) only a foolish government would guarantee that it could defend private systems from cyberattacks; (2) faith in the U.S. government rose after the September 11th attacks, largely because, in times of crisis, people need to have faith in the government; and (3) advanced societies can function quite well, even when large majorities have no faith in the people who happen to run the government. Perhaps a given attack was meant to distract the target government, hindering its ability to respond or even detect a looming challenge. Yet such an attack assumes that its target is (1) easily distracted, (2) incapable of making decisions when distracted, (3) unwilling to delegate decisions under such circumstances, (4) has bureaucracies unable to function without central direction, and (5) will not refocus on the attacker when the proximate crisis passes. This motive is particularly

---

[20] A fiendish variant is to attack computers that control manufacturing processes to retard the production of, ruin, or render dangerous the products of the processes. Such an attack could have nasty echoes. It is not clear, however, why any manufacturing process should be exposed to the outside world without very high levels of network protection.

harebrained when the crisis is military. A state's ability to wage war has everything to do with its extant military and, secondarily, its ability to mobilize its economy to resupply, rebuild, and expand its military. A cyberattack on society, however painful, affects neither very much these days. Because warfare tends to concentrate the mind once hostilities begin or even appear imminent, the effects of a pure cyberattack are likely to recede rapidly in importance. Conversely, a major cyberattack, if traced back, is certain to focus the victim's attention on the attacker, thereby obviating any advantage of surprise that the attacker may hope to profit from.

The last distraction-based motive is to raise a false flag. The attacker wants the target to retaliate (and not necessarily in cyberspace) against the supposed attacker. The real attacker may count on the target retaliating against the "usual suspects" or may shape the attack to point toward one particular state. Such an attack requires the target to recognize that it has, in fact, been attacked. This attacker must worry not only that it will be identified as the attacker, thereby making one enemy, but also that its full motive may be discerned, thereby making a second enemy (the supposed collector) as well. The only out for the attacker would occur if the target retaliates against the supposed attacker before discovering the ruse. At that point, the target may be tempted to ignore or downplay subsequent information that suggests that it had retaliated in error. By then, it may be too late; the target will already be at odds with the supposed attacker. Even if it finds the true attacker, it may be loathe to retaliate against it (at least in any visible way) because it would thereby suffer great embarrassment from admitting that its first retaliation was a blunder. Revelation alone might teach the attacker (and whoever is watching) not to make such attacks. Retaliation at that point may be icing on the cake.

Completeness suggests that criminality and "hacktivism" should not be completely ignored as motives, but the motives of individuals do not necessarily transfer readily to states. A state could, for instance, attack a bank's computers to transfer money from the its accounts into the state's own exchequer. However, states, unlike individuals looking for a big score, have a great deal to lose if they are caught—and far better ways of raising revenue. Less implausibly, attacks could come from criminal elements protected by government officials for one of many reasons (e.g., solidarity, blackmail, kickbacks), but such cases lend themselves much more naturally to prosecution rather than retaliation. Similarly, hacktivism usually involves inserting propaganda into cyberspace via site hacking or message pushing.[21] Such intrusions are of minor import; states, being states, have less-contentious ways of making their point.

Implications

Is it possible to discern motives? Sometimes, the breadth, scale, sophistication, persistence, and consistency of an MO provide good hints. Small attacks, for instance, are unlikely to be carried out for coercion. Large attacks are unlikely to be accidental. Certain classes of attacks, by definition, affect only the military. Attacks that are hard to discover are usually not meant to elicit retaliation or to impress third parties. Context, not least of which is the state of world tensions at the time, may provide a clue. In some cases, the attacker may make it fairly clear who is attacking and why. But in the end, these are all hints, and retaliation is likely to proceed under a thick fog of doubt about why as well as about who.

Nevertheless, the attempt is worth making. Motive matters in determining what strategy any retaliation is trying to frustrate. It may help indicate how much gain accrues to the attacker—not much, for many of the motives—and thus what level of retaliation is likely to dissuade mischief. Finally, motive may provide clues about how the attacker will receive any retaliation.

---

[21] If the medium of hacking is the message that a state's infrastructure is poorly protected, taking the hit and taking the hint would seem to be sounder than hitting back.

CHAPTER FIVE

A Strategy of Response

> Future administrations will have to consider new declaratory pol-
> icies about what level of cyberattack might be considered an act of
> war, and what type of military response is appropriate.
>
> *Robert Gates, Secretary of Defense, 2008*[1]

A state that has announced an unambiguous deterrence policy leaves itself little maneuvering room once another state has attacked it in cyberspace. States, however, may not necessarily want to find themselves boxed in and may thus want to explore less-risky or less-harmful options to achieve a relative degree of peace in cyberspace.

Those who decide on retaliation, if their decisions are not prepackaged, need to ponder certain choices. How public or private should the confrontation be? When should it happen? What should the state do about hackers that might enjoy the sanction of other states? To what extent should it take up the burden of retaliating for attacks against others? How effective are alternative responses that do not involve retaliation? This chapter addresses these question, then looks at deterrence from the attacker's perspective and briefly discusses signaling. It concludes with summary observations on cyberdeterrence drawn from this and the prior two chapters.

---

[1] Robert Gates, "Nuclear Weapons and Deterrence in the 21st Century," address to the Carnegie Endowment for International Peace: October 28, 2008.

Again, the lesson is that many other considerations must mediate the straightforward connection between attack and retaliation—critical to the logic of deterrence.

Should the Target Reveal the Cyberattack?

The likelihood that an attack is visible is the likelihood that the effects of an attack are visible multiplied by the likelihood that these effects will be publicly ascribed to a cyberattack (rather than to error, accident, or bad design). Both parts of the equation are far from certain. CNE is rarely apparent until an investigation reveals it. Corruption may go unnoticed until it reveals itself as a discrepancy between what a system is doing and what it should be doing. Sometimes even disruption may go unnoticed; for example, if a sensor is silent, is it silent because it has nothing to report, or has someone tampered with its reporting channel? If not people but machines or other processes consume certain services, their loss may be noticed only when the processes they feed behave incorrectly.

Normally, full disclosure is the best policy. It is too easy for governments to believe they can control information much better than they actually succeed in doing.[2] Post hoc revelation eats at government credibility—not to mention competence, if playing catch-up with events makes the government look bad. Revelation is necessary if the target state is going to respond visibly, either with retaliation or without it (using legal, diplomatic, or economic measures, for example). Going public provides an opportunity to be clear about the aims of the response. Incidentally, revelation may also be necessary for sub-rosa retaliation[3]: Just because the retaliator did not want to make a fuss about how it hit back does not mean that the attacker (as target of

---

[2]   Chernobyl is a good example; see, for example, Michael D. Lemonick, "The Chernobyl Cover-Up," *Time*, November 13, 1989, and "Protests Grow Over Chernobyl 'Cover-Up,'" *New Scientist*, October 28, 1989. China's attempt to build a huge petrochemical complex in Xiamen is another example; see Datong Li, "Xiamen: The Triumph of Public Will?" *openDemocracy*, January 16, 2008.

[3]   Recall that *sub-rosa* means *secretive* or *private*—thus out of the public eye.

retaliation) will do likewise, so it is better to get the reason for retaliation out before the fact of retaliation is revealed.

Yet silence may still be golden. Revelation may expose the fecklessness of the target's system security, reducing the public confidence in it and making it a target for repeat attacks.[4] Evidence to support the attack claim may reveal sensitive information about system security.

When Should Attribution Be Announced?

If the target does not follow up its claim of an attack with an attribution, it raises the difficult "why not" question and encourages freelancers to make up their own minds on the matter (and perhaps take independent action). Revelation may pressure the target government to announce an "arrest." If there is an attribution but it lacks thorough backup, the alleged attacker may accuse the target of slander.

Once it has revealed the attribution, the potential retaliator may have to resist pressure from its public to retaliate and from others (often foreigners) not to. Both reduce the government's ability to act according to its own best strategic interests.[5] The standard Israeli approach—"at a time and place of our choosing"—successfully allows the target time to retaliate only if the people believe the statement is not a fudge. Israel has repeatedly proved that it responds to physical attack and prov-

---

[4] A case for discretion comes from the public's tendency to overestimate the risks of cyberinsecurity; there is considerable agreement that the public is wildly inconsistent in how it reacts to low-probability, high-impact risks. These days, impressively large data breaches involving personal information are widely reported in the press, to general consternation. Less widely reported is that only a small fraction of all records so breached are associated with any fraud. In 2005, hackers compromised 163,000 customer records kept by ChoicePoint, a data-aggregator, but only 800 cases of identity theft (Federal Trade Commission, "ChoicePoint Settles Data Security Breach Charges; to Pay $10 Million in Civil Penalties, $5 Million for Consumer Redress," press release, January 28, 2006). The conversion from theft to fraud was smaller for the 2008 breach of the Hannaford Bros. supermarket chain: 4.2 million credit and debit card numbers exposed and 1,800 cases of fraud (Ross Kerber, "Grocer Hannaford Hit by Computer Breach," *The Boston Globe*, March 18, 2008).

[5] We assume the potential retaliator is the target. A later section, Extended Deterrence, moots the possibility that the two are different.

ocation. No one has that record in cyberspace. Thus, in the absence of such credibility, some sort of decision on retaliation may have to follow quickly once attribution is announced. The longer the gap between attribution and retaliation, the more time the forewarned attacker has to prepare for a return blow.[6] If retaliation is delayed, third parties (e.g., superpatriot hackers) may take matters into their own hands (but they may do so in any case, even afterward). With third-party retaliation, the target loses some control over events, but the attacker cannot use the threat of counterretaliation to avoid paying some price.

Should Cyberretaliation Be Obvious?

In cyberspace, the target can hit back against the attacker, and no one (aside from the security establishments on either side) need be the wiser. This sort of sub-rosa retaliation tends to make more sense if the attack is not public or if public attribution is not viable. In the latter case, the evidence behind attribution may be of the sort that is not easily released or not easily argued if released. Sub-rosa retaliation avoids having to make the choice of what to reveal.

States that would employ sub-rosa retaliation have to manage the expectations of those who are looking for revenge. Retaliation could still convey the target's displeasure over the attacker's leadership and could change the latter's calculus to discourage further attacks. Furthermore, the attacker, as the victim of retaliation, would not be under subsequent public pressure to counterretaliate and could therefore conclude that letting things drop after the retaliation is wiser than carrying on.

Sub-rosa retaliation, however, may be too seductive, particularly if the retaliator feels no need to convince the attacker of its guilt—after all, the attacker knows that it struck first, right? One danger is that, if the intelligence or law enforcement agency does not need to worry

---

[6]   One might have thought that the defense would have been hardened before the attack took place, but not necessarily. Attackers may be surprised that they were caught. They may worry that defensive efforts will be noticed and might reveal the operation. If the attackers come from a different bureaucracy than the defenders, they may simply not care.

about defending its attribution to others, its case to national command authorities (that is, those who control the retaliation capability) may go unchallenged. The agency may thus claim its attribution is correct when the evidence suggests a higher degree of skepticism is warranted. Furthermore, a decision to retaliate sub rosa takes certain targets off the list (e.g., power plants) or at least demands they be hit in ways that do not look like a hit (which then fails to communicate displeasure). The remaining targets may be those thought to be important to the other side's intelligence and law enforcement communities but do not directly affect the public at large. Finally, the entire strategy rests on the attacker's willingness not to make a fuss: The wisdom of the strategy is hostage to the discretion of the state that (supposedly) engineered the attack in the first place.

Adding to the seductiveness is, ironically, the saving grace that a sub-rosa response to a sub-rosa attack may be possible even when confidence in the attribution of the original attack is rather weak. The attacker, knowing that it started things, will have a fairly good idea of why it suffered retaliation and may take the hint. An injured but innocent and unsuspecting party will be unaware of what may have motivated such an attack and may have no good reason to single out the errant retaliator as the source of the discomfort.

But such cleverness can backfire. If the attacker learns about the retaliation against the third-party state, it possesses a valuable piece of information—who attacked the third-party state—and may well profit from it. Telling the third-party state who started things may seem foolish, but the attacker might be able to downplay its own role and suggest that the retaliator both overreacted and was sneaky. Alternatively, the attacking state may skip the confession but pass around hints that make it easier for the innocent victim to identify the attacker (finding something is a lot easier when you know exactly what you are looking for). Or the attacking state might blackmail the retaliator, lest its actions be revealed to the innocent victim.

Furthermore, the assumption that no one in the third-party state knows about the original attack may be wrong; it is not unknown for two states with little in common but their dislike of the United States to swap intelligence.[7] The original attack might not have been so secret prior to the attack, or its existence might have been revealed after the fact. Such revelation a may be deliberate (perhaps someone in the know is bothered by the retaliation or the possibility that it was misdirected) or may simply reflect the universal difficulty of hiding secrets.

Finally, the retaliator may have overstated its ability to keep itself anonymous. A country that suffered undeserved retaliation may not be certain who did it but may have serious enough suspicions to affect its relationship with the retaliator—and if it did not know why the retaliator acted as it did, it may be angrier than if it understood that retaliator's motivation.

Is Retaliation Better Late Than Never?

Attribution may require months and years of work (this is true beyond cyberspace; a dozen years elapsed between the Lockerbie crash and a court conviction).

In the United States, successful prosecution of a crime requires the marshalling of evidence that must pass tests rooted in the Constitution and common law ("beyond a reasonable shadow of doubt"). The decision to retaliate, however, can be informed by intelligence information and does not require utter certainty. If investigators do not know who did it, however, they may also not know whether the case is a criminal matter or a military or intelligence matter. Thus, jurisdictional conflicts between the criminal and intelligence authorities, as outlined in U.S. law, could further delay attribution of a cyberattack.[8] Other issues also have to be resolved. Seeking the cooperation of the hacker's home state, for example, depends on whether or not the attack is a criminal matter. If it is, the answer would likely be yes as well. However, if the attack were state sponsored and it was desirable to keep that state in the

---

[7] Iraq and Serbia, for instance traded information on how to defeat U.S. aircraft and avoid antiradiation missiles (William B. Scott and David A. Fulghum, "Pentagon Mum About F-117 Loss," *Aviation Week and Space Technology*, Vol. 150, No. 14, April 5, 1999).

[8] The relevant material can be found in U.S. Code, Title 18, Crimes and Criminal Procedure (18 USC) and U.S. Code, Title 50, Sec. 15, National Security (50 USC 15).

dark, the answer would be, maybe not. Cyberattacks are like terrorist attacks in that regard but are clearly different from military attacks.

Is a deterrence policy even possible if retaliation has to wait that long? Much depends on how much cyberdeterrence resembles punishment as deterrence for crime. Eventual punishment for crimes is considered justified; whether the prospect of eventual punishment deters is another issue. Because criminal justice systems have hundreds of years of history, no one doubts the will of states to detect and punish long-dead crimes. In cyberspace, which has yet to see its first retaliation,[9] the principle is not so well burnished and thus may be doubted. Until retaliation actually occurs, deterrence may lack credibility. If the attack has coercive elements, the coercion may linger. Attackers, for their part, may see no reason to stop attacking. They cannot distinguish between the target's lacking a deterrence policy and its taking time to retaliate. A lack of will, a lack of capability, and a lack of pretext all explain inactivity.

By the time the opportunity for retaliation does come, it may be overtaken by events. The attacking state may be on better terms with the target (as was Libya vis-à-vis the West by 2001). Retaliation may sour a relationship that the target may come to value. The attacker's regime may have left (voluntarily or otherwise), and successor regimes may have, in effect (even if not formally, that being unlikely) disavowed earlier actions. Irrespective of regime, if too much time goes by, the attacker may consider that the rationale for retaliation was just an excuse—especially if there were no new attacks.[10] Settling old scores

---

[9] So, why is nuclear retaliation, with no more track record, credible? Indeed, its credibility has been questioned. In the late 1970s, nuclear theorists, in fact, worried that the Soviet Union could knock out the U.S. intercontinental ballistic missile force and present a fait accompli that would inhibit a U.S. nuclear response and reveal deterrence as a hollow threat. Nevertheless, the best answer to that question is that, with the consequences of retaliation so destructive, no one really wanted to test anyone else's credibility.

[10] Suing for damages is more likely to be accepted as legitimate even after a period of years. Given the ambiguities of cyberspace and the general reluctance to admit what tools the defender uses to understand what happened to its system, the resources spent arguing over what the damages were may exceed the actual level of damages in many cases. A nominal compromise in which a token payment is made may be appropriate but hardly contributes to deterrence.

is far more acceptable in criminal matters, in which there are few serious disputes about legitimacy of punishment, than it is in cyber matters or, as a general point, in state-to-state dealings, for which there is no governing authority. Thus, deterrence delayed is nearly tantamount to deterrence denied.

Figure 5.1 illustrates a decision loop based on the considerations discussed in the first half of this chapter.

Retaliating Against State-Tolerated Freelance Hackers

Whether it is right to hold a state responsible for rogue attacks (the "culpable tolerance" doctrine) is not a settled question. A state that deliberately harbored cyberattackers and shielded them from criminal enforcement should expect little sympathy if retaliated against (questions of proportionality aside).[11]

From a strategic perspective, however, if the actual attackers are working on behalf, but not under the direct command, of the "attacking" state, is retaliation wise? Would it work?

Eschewing retaliation means that the target has no deterrence policy against states that choose to carry out cyberattacks using freelancers not under their direct command. The only compensation is that freelancers are unlikely to be called on to carry out certain types of attacks; examples of these would be serious attacks on the target's military and coercion attacks when the target is being asked to do something specific (goading attacks are also probably off the table).

The "yes" answers are, not surprisingly, also problematic. If the target state wishes to create or enforce a set of global norms by retaliating or threatening to do so, the link between the attackers and the responsible government must be established. Intelligence facts unaccompanied by revelations of sources or methods are unlikely to suffice (except among those predisposed to take the target's word). Convinc-

---

[11] Although there is scant evidence that the Taliban knew of the September 11th attack, much less that it was conducted for the Taliban. However, the Taliban's refusal to help bring al Qaeda to justice provided sufficient *casus belli*, rendering irrelevant the question of original culpability.

Figure 5.1
A Decision Loop for Cyberdeterrence

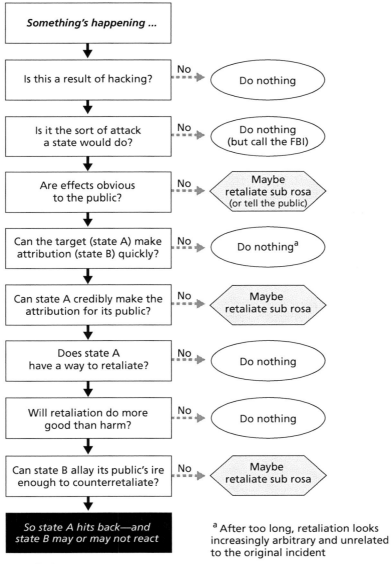

Something's happening ...

Is this a result of hacking?   No → Do nothing

Is it the sort of attack a state would do?   No → Do nothing (but call the FBI)

Are effects obvious to the public?   No → Maybe retaliate sub rosa (or tell the public)

Can the target (state A) make attribution (state B) quickly?   No → Do nothing[a]

Can state A credibly make the attribution for its public?   No → Maybe retaliate sub rosa

Does state A have a way to retaliate?   No → Do nothing

Will retaliation do more good than harm?   No → Do nothing

Can state B allay its public's ire enough to counterretaliate?   No → Maybe retaliate sub rosa

So state A hits back—and state B may or may not react

[a] After too long, retaliation looks increasingly arbitrary and unrelated to the original incident

RAND MG877-5.1

ing the attacking state to call off such attacks or prevent their taking place faces even greater difficulties when the offenders are freelancers. The logic (see below) that feigning innocence may exacerbate the target state's doubts about who did it is stronger when retaliation depends on both finding the hackers and linking them to the government.

Retaliation presumes that pressuring the attacker-government can end the activities of the freelance attackers themselves.[12] The presumption will be mostly true if the attackers were directed and compensated by the government for their efforts. The looser the connection, the looser the gearing between punishment and cessation. There is a big difference between cutting off funding to or otherwise discouraging freelancers on the one hand and discouraging their activities by, for instance, prosecuting them. It is hard to justify retaliation against a government that apparently would like to but cannot prosecute, or even find, mischief makers that other states believe are in the state (the link is tighter if the government knows who they are, knows where they are, but prevents others from taking action against them). Governments (notably in the third world) that can legally prosecute individuals may find it politically difficult to do so if the accused are influential. The distinction between will not and cannot can be very difficult for an outsider to make, much less prove. Perhaps no retaliation can effectively stop or even retard freelancers with multiple customers.

If retaliation nevertheless takes place, the attacker's government may be left with the problem of convincing the retaliator that it has done all it can to rein in its hackers. But how can a state under the threat of retaliation prove that it is going after hackers with sufficient

---

[12] Ross Anderson, who is by no means a cyberwar hawk, nevertheless argued that,

> [i]f our air-defense threat in 1987 was mainly the Russian air force, and our cyber defense threat in 2007 is mainly from a small number of Russian gangs, and they are imposing large costs on US and European Internet users and companies, then state action may be needed now as it was then. Instead of telling us to buy antivirus software, our governments could be putting greater pressure on the Russians to round up and jail their cyber gangsters. (Ross Anderson, *Security Engineering*, Indianapolis, Ind.: Wiley, 2008, p. 220.)

Elsewhere, Anderson notes that Russian police "were prodded into arresting the gang responsible" for using botnets to extort money from online bookmakers (see Anderson, 2008, pp. 198 and 640).

energy and diligence? Must it turn over all records that the injured state claims are relevant? If the accused state lacks the technical capability to find hackers, is it obliged to give personnel from the injured state carte blanche to conduct investigations within its borders?[13] If rogue employees, especially those connected to intelligence services, are under investigation, what aspect of their "work" can legitimately be kept from such investigators?

Would retaliation increase the efforts of states to ferret out the hackers within? Or would it highlight the unfairness of the exchange: You actively attacked us because of our passive failure to do everything possible to prevent an attack by others? Would the attacker then dig in its heels? If retaliation followed, would the attacker accept this as its due or strike back? The attacker's government may convince itself that the only way to relieve the pressure is to counterretaliate (ironically, justifying the retaliation, in retrospect) and hope the retaliator concludes that the costs of pursuing the attacker's government exceed the benefits accruing from the small likelihood that the pursuit might limit freelance activities. Worse, after retaliation, the original hackers may attain the mantle of heroes—they struck a state, whose subsequent cruelty gave retrospective justification to the original attack. Hezbollah emerged from its 2006 confrontation with Israel much more deeply embedded in Lebanon, even though its kidnapping of Israeli soldiers brought the wrath of outsiders down on the citizenry's heads.[14]

Note that, if the retaliator is going after states that could have stopped the attack but did not, the rationale for retaliation shifts from juridical (punishment is the legitimate consequence of crime) to the pragmatic (punishment is a useful way to get the other party to take corrective actions). The pragmatic argument does not lend itself to third-party explanation so well, unless it is possible to prove the international equivalent of gross negligence. That is a much higher stan-

---

[13] States balk at giving outsiders too much influence over how they enforce their laws. In 1914, Serbia had accepted all but one element of the ultimatum the Austro-Hungarian empire imposed after one of Serbia's citizens shot Archduke Ferdinand in Sarajevo; it resisted making foreign officials part of its own justice mechanism.

[14] Amos Harel and Avi Issacharoff, *34 Days: Israel, Hezbollah, and the War in Lebanon*, Basingstoke, UK: Palgrave-MacMillan, 2008.

dard than "not doing what you could have." For this reason, pragmatic retaliation may work best through sub-rosa means.

What About Retaliating Against CNE?

Despite the big difference between CNE and cyberattacks, there is undeniable irritation within the U.S. government over the computer intrusions (which are illegal by U.S. law) of the sort China supposedly carries out. The fact of espionage may be but one component of the ire. The volume of the intrusions (in terms of both number and bytes exfiltrated), their apparent brazenness (e.g., going after networks the Secretary of Defense uses[15]), and the mess made in the course of doing so just make things worse.

Can retaliatory cyberattacks be used to stop or at least tamp down such CNE? Anything that penalizes an activity is likely to dampen the net benefits of carrying it out. But even if attribution is correct and if retaliation hurts the spymaster without depleting the stock of cyberweapons available, the risks may argue against it.

The first hurdle is communicating exactly what was objectionable about the behavior—was it the fact of espionage, its volume, its gross nature, the mess it left behind? An ancillary hurdle, for future occasions and for third parties, is translating one instance of objectionable behavior into a broad tenet: What defines *gross* or *mess*? The easiest position is to define CNE itself as objectionable but this is the position least likely to be accorded legitimacy (because "everyone does it").

A second hurdle is determining a threshold for response. In a world in which tens of millions of computers are bots and in which no important network goes unprobed, a no-threshold policy is very hard to implement. But what level of activity would be actionable? Many estimates of the cost of purloined information tend to measure loss by the cost; for example, in the case of certain types of intellectual property, of conducting the research. Information that cost $1 million to generate, however, is not necessarily worth $1 million to the

---

[15] Demetri Sevastopluo, "Chinese Hacked into Pentagon," *FT.com*, September 3, 2007.

attacker; its possession by others, even competitors, does not necessarily equate to a $1 million loss to the victim. A how-much-did-it-cost-to-generate-the-information rule may have numbers to back it up but still be meaningless. Proportionality requires that punishment reflect how much was *actually* taken, not just how much was *discovered* as having been taken. In a realm in which very little is detected, how much can be inferred about unseen activities (which cannot be measured) from those that are seen (which can be measured) is questionable.

A third hurdle is ascertaining that an act of retaliation had any effect on the behavior of the offending state. Facing pressure, the spymaster might (1) do nothing different (while claiming it is doing nothing, doing something, or leaving the matter in doubt), (2) swear to carry on but stop anyway, (3) swear to back down and not do so, or (4) swear to back down and actually do it.

What the spymaster says will be based, in part, on what rules the spymaster wants to be in place (e.g., is CNE legitimate?) and what reputation it wants to maintain (e.g., cooperative and contrite versus defiant and unbowed). If the retaliation (both effects and retaliator) is public, the spymaster has little to lose by asking for proof; the retaliator, however, puts its sources and methods at risk by providing that proof.

What the spymaster does will be based on what it can get away with, which, in turn, is related to how good it thinks the target's attribution capabilities are and whether the spymaster can get almost as much done with fewer fingerprints. If the spymaster has been outsourcing its espionage to talented amateurs and getting into trouble for what they have done, it may want to take such efforts in house. Conversely, if the target is trumpeting discovered links into the bureaucracy, some outsourcing may be called for. The spymaster may seek an MO that is different from the one it was caught using; it may cull the less-valuable targets, the better to harvest more-valuable targets quietly. It would also tend to be tidier in general and perhaps less brazen. The only long-term effect of retaliation, therefore, may be to persuade the spymaster (1) to bring its prior efforts freshly to the attention of the state's decisionmakers, (2) to analyze what it really needs from its future efforts, and (3) to professionalize its tradecraft.

Should Deterrence Be Extended to Friends?

Should a "big retaliator" offer to retaliate against attacks on a "little target"?[16] Following the cyberattacks in 2007, Estonia wanted NATO to decide that (what was thought to be) Russia's cyberattack triggered the alliance's collective defense clause; NATO said no.[17] Would saying yes (assuming the response was in kind) have made technical and operational sense?

Arguments in favor draw on familiar lines of thought. A big retaliator may have the capacity for cyberretaliation that the little target may lack. The little target may be more vulnerable because of insufficient infrastructure redundancy. Cyberdeterrence could remove one avenue for the attacker to pressure another state's allies.

Technical issues, though, cast doubts on extended deterrence. Attribution and battle damage issues will plague extended deterrence, only more so. Perhaps the little target gives the big retaliator deep access to its systems (and no state trusts another's intelligence operatives). Perhaps the damage is obvious (as with Estonia). Otherwise, when it comes to knowing what the attack did and who did it, the big retaliator has to take the little target's word for it ("Why dig in our systems? Don't you trust us?"). In 2007, Estonia, in accusing Russia, appeared to play the role of a trustworthy and honest (if yet-to-be-proven-correct) victim. It would not be unprecedented in the annals of diplomacy for a state to fabricate or exaggerate evidence that implicated a traditional enemy for carrying out an attack that may or may not have happened and

---

[16] The question in the heading is different from asking whether one state should offer to defend the cyberspace of another state. Helping is one thing; one state can share its intelligence and its tools with another. One having the means to scour cyberspace looking for bad code can also share that capability (or, alternatively, scout and scour worldwide for all sorts of bad code, router owners permitting). But defending another state's cyberspace requires being given sufficient access to its systems. Whether or not the defending state trusts its external defender is a serious question. Installing an intrusion-detection system in someone else's system is one thing. Getting sufficient administrative privileges to root around for vulnerabilities is quite another. Private information security may be a better choice for such ticklish tasks.

[17] Ian Traynor, "Russia Accused of Unleashing Cyberwar to Disable Estonia," *The Guardian*, May 17, 2007.

thereby caused damage that may or may not have crossed a threshold. Even if the big retaliator exercises due diligence and demands to see the evidence, who knows whether it has been denied access to exculpatory evidence?

Coordinating retaliation is another challenge unless the little target leaves the whole matter to the big retaliator. Deconflicting targets is straightforward so long as there is no one target that both want to hit because it is particularly appropriate to the nature of the attack. Going after the same target may require one or both sides to share their own tricks with the other, something each may balk at doing. Otherwise, clumsy attacks by one may alert the original attacker in ways that frustrate sophisticated attacks by the other.

Extended deterrence can also be offered anonymously—why not? Cyberdeterrence prevails if attackers believe that an attack on *anyone*, rather than a specific someone, will have deleterious consequences regardless of who attacks whom (recall the discussion about third parties above). Thus, if *some* states were willing to retaliate for attacks on other states, this could simultaneously create a broader deterrence without necessarily implicating the true retaliator—unless only one state steps up to the role (while the target conspicuously does not).

One caveat: If the nature and effects of the original attack were not of the sort that would be universally visible, the attacker may feel, in the wake of retaliation, that the target state was culpable even if it did not carry out the retaliation itself. The target's revealing enough information to a third party to enable retaliation may suffice to justify counterretaliation. If the target can be made to believe as much, it may keep quiet even to its friends.

Finally, there is something to be said for extended defense. Observe that, in 2008, Georgia evaded DDOS attacks by rehosting its Web sites on U.S. servers with capacious fiber optic connections and adroit system managers.[18] Although Georgia's physical geography made it vulnerable to Russian pressure, its emerging cybergeography was quite compatible with U.S. friendship. Should other states that are

---

[18] Stephen Korns, "Botnets Outmaneuvered," *Armed Forces Journal*, January 2009, pp. 26–28, 38–39.

not bound in U.S. alliances be attacked, the United States can still be there to help. Enough of this may convince attackers that their efforts, while tactically sound, are strategically counterproductive.

Nevertheless, DDOS attacks are the previously noted exception to the rule that attacks stem from vulnerabilities. Otherwise, there is little ipso facto reason that a small state cannot defend its networks from attack as well as a large state can. Line for line, the code its networks run is no more likely to be buggy than the code that networks in large states run. Indeed, in some ways, because smaller networks tend to be less complex and because personal trust relationships are easier to define and monitor, defense is easier than it is for large networks. Thus, the rationale for collective action that impelled the formation of NATO in 1948 does not exist in nearly the same degree in cyberspace.

Should a Deterrence Policy Be Explicit?

Everyone is well aware that any hostile and damaging act directed against the United States (or any other similarly equipped state) may call woe on its head. Whether or not the exact nature of the hostile act has been called out is entirely secondary. To demonstrate, consider this (admittedly far-fetched) tale. A nasty state inserts subliminal messages into major network broadcast feeds that persuade many Americans to sabotage the transportation system. Thousands die. The United States retaliates in ways that leave no doubt why. The attacker screams, "unfair!" because it no one never made it explicit that using subliminal sabotage messages was the kind of attack that the United States would retaliate against. Any fair-minded person would laugh at such protests. Ditto for cyberspace: Hostile intent, illicit behavior, and consequent damage suffice to justify retaliation, whether or not the cyberattacks, per se, are included.

For deterrence to hold, every potential attacker must believe that retaliation may follow from any hostile attack. No sane state would promise that retaliation *will* follow from a hostile cyberattack because the problems of attribution are so difficult. This is why an explicit deterrence strategy may hurt more than help. True, an announcement would make retaliation more credible by making it harder for the United States not to respond to a bald-faced attack. Still, coupling a declaration with an uncertain response capability could, if put to the test, reduce the credibility of all *other* forms of deterrence (e.g., the threat of air strikes). Fear of losing related credibility would pressure states to respond even when ambiguities remained. The forensic (intelligence and law enforcement) community may be tempted to offer psychological assurances from which qualifying doubt had been removed. The strategic community, which may care little about what happens in or because of cyberspace, may be painfully aware that the credibility of deterrence is an indivisible swatch in the fabric of statecraft—and may thus counsel badly.

Any declaration will itself cause potential attackers to ask themselves: Why this; why now? Perhaps they will conclude that the declaration is driven only by internal bureaucratic politics. But if potential attackers infer that a deterrence policy is needed because implicit postures are inadequate, they may conclude that the declaring state has discovered (1) that its cyberdefenses could no longer be trusted or (2) that it has more to lose in a cyberattack than one might have guessed. If cyberretaliation is promised for cyberattack, the attacker may begin to wonder whether noncyber means have been taken off the table.

One alternative to an explicit policy is to wait for easy-to-attribute attacks; retaliate in kind; and, if these retaliatory attacks succeed, retroactively justify them. Others could infer a deterrence policy from such action. The advantage of waiting for good attribution and reliable BDA (confirming the ability to retaliate) is that, when the policy is announced, few will raise issues about attribution and BDA in the face of an existence proof. Furthermore, a threshold will have been demonstrated. If no counterretaliation ensues, only continuation, the failure to disarm, and fortification would continue to be barriers—perhaps at that point, speed bumps—to the widespread adoption of cyberdeterrence as a policy.

Conversely, a one-time response does not make a law universal. Those contemplating attack may figure that they will not be caught and that their defenses will resist retaliation. The retaliator may be perceived as having one rule for the weak and feckless and another for the strong and serious. Furthermore, although success at retaliation establishes deterrence, counterretaliation or escalation would push the participants into a cyberwar or worse as the attacker tries to do by action (responding to a small but obscure attack) what it might have been able to do with mere words (an explicit declaration). While waiting for the right pretext, the retaliator also lacks anything but an implied deterrence posture. Ironically, if the goal is to establish a policy of deterrence by action, there is no requirement that retaliation actually force the other side to stop. But if it does not, the inescapable conclusion is that the deterrence policy, when activated, did not, in fact, deter.

The United States is no stranger to ambiguous deterrence policies. The United States wishes to make it clear to the Chinese that their invading Taiwan might launch them into a war with the United States in short order. But the United States also wants to leave some question in Taiwan's calculus about whether it might actually come to Taiwan's aid, lest it pocket the U.S. guarantee and then declare independence, which would force the United States into a confrontation with China. Similarly, it was never clear whether or not the United States would respond with nuclear weapons to a Soviet conventional attack on Europe. In both cases, ambiguous deterrence has worked so far—which is to say, at least in the first case, it has yet to fail.

Appendix B discusses a model for comparing the costs and benefits of explicit and implicit deterrence postures depending on how the posture affects the likelihood and the consequences (outcome-value) of a state attack, a target response, and attribution error.

Can Insouciance Defeat the Attacker's Strategy?

If cyberattacks are being employed coercively, defeating the attacker's strategy may mean doing nothing.[19] This may signal that the cyberattack did not hurt or (if success is not obvious) even register.

---

[19] Today's adversaries have multiple ways of making their opponents' lives miserable, so it pays to persuade adversaries to invest their efforts in endeavors that consume their resources but give you little pain or that are least likely to defeat your strategy. If a state deems cyberattacks the next best thing to futility, it might want to convey pain in a way that precisely

Doing nothing is particularly attractive if cyberattacks were used to goad the target into action. A strong reaction, conversely, may tell the attacker that the cyberattack hurt enough to persuade the target to take risks to stop further pain. An attacker that hears a yelp of pain may figure that it is close to achieving its goals—so long as it continues its cyberattacks. If private systems have been attacked, insouciance says that the government does not feel it has a responsibility to stop or even moderate the pain of feckless system owners. Insouciance declares, go ahead, work yourself up, attack us, get a reputation as a bully; we are so big that we notice little and care less (and anyway, it may be someone else's problem).

How the *attacker* responds to the target's insouciance will depend on whether the attack was its best punch. If so, responding by repetition is likely to make less of an impression and is thus probably a weak approach. The attacker could give up, all the while denying (if necessary) the fact, the authorship, or the intent of the attack. If the attacker had a better punch, it could try to land it and see if that would have any effect. In the latter case, the target's insouciance strategy might have backfired, and the time may come to suggest that it has a retaliation capability that it is ready and willing to use, if matters so dictate.

Maintaining insouciance (if the attack is revealed) requires ignoring political calls for action—which perhaps takes only a modicum of courage. Cyberattacks rarely hurt anyone or break anything, so calls for action are unlikely to have the emotional resonance of the sort engendered by bloody terrorist attacks. If the target system is private (which it would practically have to be to disrupt life), the argument that victims were feckless further insulates the government from having to react.

Confrontation Without Retaliation

Highlighting and attributing attacks, at a minimum, provides one's sysadmins a vivid illustration of which potential attacks to watch out for and from which sources. If no further action is taken, all the risks

---

diverts attackers from more cost-effective mischief into what might be more-expensive, but futile, cybermischief.

and complications of retaliation are avoided. Unfortunately, so are the benefits.

Many of the motives for cyberattack are likely to shrivel in the sunlight of disclosure. If the attacker thought the target attacked first, disclosure could help clear up misunderstandings. Disclosure alone might persuade the attacker to tighten control over its rogue operatives. Attacks meant to boost the attacker's prestige relative to the target's would likely backfire if revealed. The same holds with more force if the attacks were designed to get a third party in trouble. If the motive was weak but the fear of disclosure was great, the attacker should be disabused of the attractiveness of starting trouble.

Yet nothing is guaranteed. The attacker may deny and keep on denying its involvement in the face of what a neutral party would consider damning evidence. It may demand evidence of the sort that forces the target to refuse to hand it over (thereby casting doubt on previous evidence) or to reveal forensic sources and methods. In such behavior, the attacker may find another justification to continue attacks. If serious purposes, such as coercion or attempts to damage armed forces, are involved, mere disclosure may not do much more than cause temporary embarrassment. If the hostility is so great that the attacker will take pains to give pain, mere words will likely not do. Overall, if revelation suffices, the risks of going further need not be borne.

Targeted states have other ways to communicate displeasure. The case for prosecuting individuals may be good if the attacking state cooperates (e.g., because it was a rogue attacker).[20] Individuals, even those in official positions asked to carry out implied instructions, may be deterred the next time; ditto for higher-ups. Attack methods may be detailed. If the attacking state balks, perhaps for such reasons, its lack of cooperation provides an opportunity for the target state to press the matter to convince others that the attacker has something to hide.[21]

---

[20] *May* if the evidence required to finger an individual within a state exceeds what is required to assign responsibility to a state. Responsibility can entail sins of omission, as well as commission.

[21] A great deal will depend on the particulars and the cogency of the attacker's objections. Not all of them would be specious—the U.S. Constitution, for instance, would prevent foreign investigators from using techniques here that may be routine in their own nations.

How hard to push is an open question. What would be gained by continuing to show the attacker in a bad light on this issue? Can the rest of the world be convinced to pay attention and align against the attacker, or will the target state get the reputation of being a hapless whiner, seeking to make a something out of virtual nothings? The ability to garner allies directly affects the efficacy of a diplomatic or economic response. Unilateral restrictions, even by the United States, may hurt the one doing the imposing more than the one being imposed upon.

Limiting the attacker's use of cyberspace may be poetic judgment of sorts but pressing the world community to cut the attacker off completely would probably be impossible and might backfire (e.g., if isolation makes the attacker behave worse).[22] A more-tractable approach may be getting the world community to structure Internet and communications channels in a way that gives the attacker nonprivileged positions (e.g., its router connections are paltry, its information technology firms get little business, third-party traffic cross-hauling arrangements are disadvantageous).

It bears remembering that, in deciding whether to press the issue or in persuading others to help press the issue, a given state's position on another state's behavior in cyberspace is only one aspect of its broader posture vis-à-vis that state. Whether one state chooses to press another on its displeasure would depend on what else is on its agenda. Take U.S.-Chinese relationships. The United States simultaneously wants the Chinese to help it keep North Korea under control, price the Chinese currency fairly, buy U.S. Treasury bills, and reduce carbon emissions. If the United States were to add "behave nicely in cyberspace" to that list and could get past the expected denials ("we *do* behave nicely"), it might have to decide what it was willing to give up to reach a mutual accommodation. Would it put more on the table (i.e., do more of what China wants the United States to do) or cut back on U.S. demands? If the United States wants it all, can it expect U.S.

---

[22] Senator John McCain, for instance, talked of cutting Russia out of the Group of Eight for its growing belligerence in general and its role in the attacks or cyberattacks against Estonia in particular. See Eli Lake, "McCain Backs Tougher Line Against Russia," *The Sun* (New York), March 27, 2008.

allies to join in pressuring China, when their agendas might be completely different? With a state whose cooperation in other matters is less critical (say, Russia, for argument's sake; it does not buy so many U.S. Treasury bills), enforcing good conduct in cyberspace may come higher on the U.S. agenda. The bargaining leverage may come from being more willing to apply pressure.

Finally, if the attacker is a state the United States does not much like (say, Iran), a U.S. demand that it behave nicely in cyberspace is just one more stick to beat it over the head with (but even with Iran, the United States may have higher priorities). None of these calculations would be necessary if all states honestly professed to want good conduct all around in cyberspace and saw mutual interest in suppressing cybermischief, regardless of its source. The big states, however, are more likely to give themselves a pass in this regard, even as they press others.

The Attacker's Perspective

Whatever credibility problems a cyberdeterrence policy may have from the target's perspective, it may still look credible enough to the attacker to inhibit cyberattacks—which is what such a policy is supposed to do. With or without an explicit deterrence policy in force, an attacking state has to factor in some likelihood that an attack will engender a response. The questions are—how likely and how bad?

Any state contemplating attack can be expected to weigh not only the odds of immediate success but also the odds that it can avoid punishment. The odds depend on the likelihood that the target (or, less likely, a third state with an interest in stopping cyberattacks) will

1. detect the attacks
2. identify the attacker correctly, with an actionable level of confidence
3. judge that these attacks crossed some threshold
4. strike back
5. cause real pain.

All five must be true before deterrence is in play. The odds that the target will strike back in turn may reflect

1. whether it can wreak enough damage to get the attacker's attention
2. how much it fears counterretaliation
3. how much control the target can maintain over the action-reaction cycle (especially vis-à-vis third parties).

Thus, what matters is the target's forensics capability (which also helps keep third-party actions from confusing matters), retaliatory capacity (judged against the attacker's own vulnerabilities and defenses), and its overall attitude (e.g., bloody mindedness with a high tolerance for mistakes in attribution and targeting). Bear in mind that a calculating attacker has already factored in some expected level of retaliation and has attacked anyway. Anything the target does or says will be measured against such expectations. Since neither attitude nor capabilities in cyberspace can be validated directly, the attacker has only words to go by. Words are evidence of posturing; they are imperfect guides to action.

Finally, fingering the attacker and lining it up for retaliation may persuade it to continue attacking regardless of its diminishing returns from doing so. If the actual attacker can introduce doubts into the minds of the target's leadership that the attribution is mistaken, the leadership might reconsider its decision to keep hitting back (as well as question whether it occupies the moral high ground). Now, assume a nonzero a priori doubt in the target's mind and let Bayesian logic take over. If the target is correct, the attacker may respond to retaliation (that exceeds prior expectations) by rethinking the logic of cyberattack and backing off. If the target is incorrect and punishment is misdirected, the real attacker (who escaped punishment) has been given no reason to back off, and the presumed attacker cannot back off because it was not doing anything in the first place.

Thus, if attacks continue after retaliation commences, the a posteriori position that attribution is correct becomes less likely (because having little to no effect is only one of two possibilities), while the a posteriori position that attribution is incorrect is unaffected (because having little discernable effect on future attacks is by far the most likely possibility if the true attacker was not affected). The probability needle swings toward error. The failure of retaliation to make much difference may thus shake the target's confidence in its own powers of attribution. If this logic is complete, the attacker's strategy is to act as though it had nothing to do with the original attack.[23] Since the attacker knows where the real evidence lies, it can feign a limited degree of openness to investigation. Meanwhile attacks continue.

Signaling to a Close

Deterrence is a form of signaling.[24] It communicates to potential attackers that attacks will be met with retaliation and that forbearance means being left alone. The deterring state does so in the hope that attackers weigh the consequences of attack in their calculus to make attacks. But cyberspace is very noisy; neither attacker nor target can be sure of what happened or who was responsible.[25]

The Israeli-Palestinian dynamic is an object lesson in how easily signals get lost. In theory, Israeli policy is to retaliate following acts of terrorism. This would signal to Palestinians in general, and Hamas in

---

[23] The relationship between retaliation and continuation may not be so clear-cut. Many other factors besides retaliation affect whether cyberattacks continue. For example, they may continue but with less effect, so no one notices the later attacks. If retaliation is visible and if the presumed attacker has been falsely identified, the real attacker might halt attacks to validate the target's initial erroneous attribution. Finally, if the target anticipates the attacker's strategy, such a strategy will be discounted—and so on down the hall of mirrors.

[24] See Thomas C. Schelling, *Arms and Influence*, New Haven: Yale University Press, 1966.

[25] In 1962 (the Cuban Missile Crisis) and 1973 (in the latter weeks of the Arab-Israeli war), U.S. nuclear forces were put on higher alert to indicate to the Soviet Union how serious the United States was about the crisis at hand and its willingness to use nuclear weapons, if necessary. There may not be any useful counterpart to such signaling in cyberspace. As argued in Chapter Four, issues of will are less salient in the cyberrealm, where the real question becomes that of capability—which, to a large extent, is a question of whether the other side has the requisite vulnerability. Credibly signaling that the other side has a vulnerability, however, is self-defeating if it can then detect and close that vulnerability.

particular, that terrorism is costly. In contrast to cyberattacks, terrorism in Israel used to be fairly straightforward. Attribution was easy: It was almost always a Palestinian, and, even if not, they cheered anyway. BDA was obvious. Creating a threshold—loss of life—was clear and easy to measure: Was someone hurt? Retaliation consisted of killing leaders or key warriors of Hamas (or Fatah, Islamic Jihad, et al.). But the retaliation cycle was not particularly clean. One problem, ironically, was that the Israelis believed that retaliation could, in fact, disarm the group. If Israel saw a good opportunity to strike back, it was likely to do so, whether or not such a strike could be tagged to a prior Palestinian attack—so its foes probably figured that it was going to be targeted if it was vulnerable, whether or not it had just done something specific. Hamas, for its turn, discovered what a bloody nuisance could be created by firing rockets across the Gaza-Israel border. This created a threshold problem for Israel. Since it has not responded to every attack (the vast majority do nothing[26]), Israel could not begin to do so without its reaction being widely viewed as highly disproportionate. No number (e.g., retaliate after ten attacks in a day) would be seen as anything but an arbitrary threshold. Retaliating after Hamas got "lucky" and caused horrific damages (e.g., by hitting a kindergarten) would convert deterrence policy into a game of Russian roulette. Eventually, Israel, having been fed up, struck back in January 2009, but the lack of a temporal relationship between the rockets and the reaction fed suspicions of ulterior motives—and all this for a case with few ambiguities.[27]

Because cyberspace is noisy, both the easily understood and subtle signals (thought to be) present in the nuclear realm may be nearly indecipherable in the new medium. Noise destroys communication, hence signaling. Without clear signaling, it is difficult to distinguish deterrence from aggression. Because attribution is so difficult, the only unambiguous signal would be from an attacker that identifies itself,

---

[26] As of May 2008, over 3,000 Kassam rockets had been launched, and 15 people had been killed (Israel Ministry of Foreign Affairs, "The Hamas Terror War Against Israel," Web site, August 3, 2008).

[27] These events took place just before the Bush administration left office but a month or two before the Israeli elections.

either explicitly or by unmistakable association (e.g., if air defense computers suddenly went dark minutes before a sneak attack). In peacetime, attribution requires an adversary that virtually dares deterrence.

So should the United States (or any state for that matter) have a deterrence policy? The best answer may Zen-like: No state should have a deterrence policy, but neither should any state foreswear retaliation. No attacker should conclude that a successful attack that goes unanswered means that the target state was incompetent (unable to attribute or unable to land a blow after attribution) or was bluffing in its threats. Every attacker should fear that, if its efforts succeed too well, retribution might just follow. If there is to be cyberdeterrence, it must be a policy that leaves more than just a little to chance.

Although the most important factor in weighing retaliation is attribution, a close second is whether or not the effects of the attack are public. If the attacks are public, the dynamics of retaliation will be determined by the fact that everyone is watching. Whether and how to respond in cyberspace are likely to reflect the face a state wishes to show in the arena. If the attack is private, saving face may not matter—the contest is between the leaderships of two states only. In such a case, a targeting philosophy that communicates displeasure and resolve to the opposite leadership, without either losing face, is probably the least bad way of escaping what would otherwise be a messy and potentially dangerous situation.

CHAPTER SIX

Strategic Cyberwar

A campaign of cyberattacks launched by one entity against a state and its society, primarily but not exclusively for the purpose of affecting the target state's behavior, would be *strategic cyberwar*.[1]

The attacking entity can be a state or a nonstate actor. To echo the distinction made in Chapter Two, if the attacker is a nonstate entity, it is unlikely to present much of a target for the defending state to hit back against, although states that support or tolerate them may be subject to countercoercion. For this reason, this discussion will focus on state-on-state contests, which present a richer set of questions.

Here, one assumption is that no other active hostilities are taking place (or if they are, are clearly secondary to cyberwar). This distinguishes strategic from operational cyberwar: the use of a computer network attack to support physical military operations. A second assumption is that cyberwar is two sided (although a one-sided campaign is certainly possible). The final assumption is that resorting to cyberattack and cyber counterattack means that legal, diplomatic, or economic responses to cyberattack have not come into play or have been deemed insufficient to forestall recourse to mutual retaliation.

This chapter first reviews the purpose, then the limits of cyberwar. It next moots the advisability of conducting cyberwar sub rosa and what the government can do in defending against cyberwar. The

---

[1] In this context, two *nonstate* actors would wage cyberwar on one another, but the means often employed—hacking each others' Web sites—are almost beneath serious notice. Something of this sort took place in the wake of the 2001 Hainan Island incident involving a confrontation between a Chinese fighter aircraft and a U.S. P-3 surveillance plane.

final topic is the conduct of cyberwar and the potential paths from cyberwar back to cyberpeace.

Overall, very little in this chapter supports the wisdom of cyberwar, except perhaps as a means of persuading others of its inadvisability.

The Purpose of Cyberwar

States could find themselves at cyberwar in one of two ways: through deliberate provocation or through escalation. A cyberwar could arise deliberately, from one state's belief that it can gain advantages over another by disrupting or confusing the latter's information systems (akin to strategic air attacks in World War II). A cyberwar might also start as escalation and counterescalation in a crisis take on lives of their own (more akin to the mobilization contest of World War I). In either case, the onset of cyberwar means that primary deterrence has failed. That noted, however, secondary deterrence—the ability to establish do-not-cross lines—may still succeed.[2]

In either case, it must be assumed that states engage in cyberwar to accomplish certain ends, rather than as an end in itself. True, it cannot be presumed that states are entirely rational actors in the sense that they assess the costs and gains dispassionately. Many a war has dragged on because belligerents feared that, irrespective of tangible gains or losses, the first to withdraw from combat would lose face and find itself subject to another's agenda. Such motives could well pervade cyberwar. It is merely necessary to assume that *some* degree of instrumental rationality persists.

Cyberwar has external and internal objectives. The external objective is the reason for cyberwar in the first place (e.g., to bend the other side to one's will). The internal objective relates to managing the fighting itself (stopping it, limiting its scope as, for instance, by not putting life-protecting systems in play, etc.), and avoiding escalation into violence. Note that, as with other forms of war, the two objectives may

---

[2] Potential examples of secondary deterrence are the successful dissuasion of Nazi Germany and, half a century later, Saddam's Iraq from using chemical weapons.

appear to be in conflict. Teaching someone a lesson that cyberviolence does not pay by hitting them back harder calls for more aggression; inducing them to reduce their aggression by showing forbearance calls for a calibrated reduction in aggression. One objective that cyberwar *cannot* have is to disarm, much less destroy, the enemy. In the absence of physical combat, cyberwar cannot lead to the occupation of territory. It is almost inconceivable that a sufficiently vigorous cyberwar can overthrow the adversary's government and replace it with a more pliable one.[3] Ultimately, because cyberwar cannot disarm cyberwarriors, the contest becomes, as the Duke of Wellington is reported as saying, a matter of "who can pound the longest" and takes it better.[4] Perhaps neither can pound long enough or hard enough to do anything more than annoy the other.

Just because cyberdeterrence is problematic does not mean that cyberwar is implausible. Table 6.1 suggests that some of the factors that work against cyberdeterrence do not apply to cyberwar. Attribution and thresholds cease to become issues once states commit to carrying out extensive cyberattacks against each other, especially if in retaliation. Similarly, since deterrence has failed, it is moot whether or not a deterrence policy creates a moral hazard.[5] Escalation is still a problem, although the longer the cyberwar continues without escalation, the likelier it is to continue that way. Incomparability of targets may alter the calculus of combat, but whoever is carrying on the conflict must have found an interesting target set. Similarly, although emerging hackers can blur the signals, the fact of cyberwar may be the only message that

---

[3] See the next section, "The Plausibility of Cyberwar."

[4] Schelling, 1966, p. 201, has described such a conflict (admittedly at the nuclear level) as

> a war of pure coercion, each side restrained by apprehension of the other's response. It is a war of pure pain; neither gains for the pain it inflicts, but inflicts it to show more pain can come. It would be a war of punishment, of demonstration, of threat, of dare and challenge. Resolution, bravery, and genuine obstinacy would not necessarily win the contest. An enemy's *belief* in one's obstinacy might persuade him to quit. But since recognized obstinacy would be an advantage, displays or pretenses of obstinacy would be suspect. We are talking about a bargaining process, and no mathematical equation will predict the outcome.

[5] At that point, explicit indemnification policies—which could go either way—will influence what private institutions spend for cyberdefense.

Table 6.1
Not All Factors That Make Cyberdeterrence Problematic Make Cyberwar Problematic

Question	Effect on Cyberdeterrence	Effect on Cyberwar
Who did it?	Cannot know whom to retaliate against	Target has already been selected for other reasons
Hold assets at risk?	Do not know whether retaliation will have desired effect and thereby deter	Important, but not critical, to know effects to justify and shape further effort
Repeatedly?	Cannot know whether retaliation is repeatable	Will affect intensity of effort over time
Disarm?	No second prize (e.g., disarming) for failure to deter	Cyber war can only be countervalue not counterforce
Third parties?	Will interfere with signaling	Signaling not important, but may interfere with escalation management
The right message?	Deterrence policy may create moral hazard	Moral burden has already been accepted
Threshold?	Will interfere with signaling	More-important thresholds have already been crossed
Avoid escalation?	Risks of counterretaliation may reduce credibility of retaliation	Physical war is already more escalatory than cyberwar
Little worth hitting?	Retaliation could be an exercise in futility	If a cyberwar ensues, it can only be against a state with a good set of targets

needs to get through anyway. BDA, in terms of what one can threaten, is irrelevant once threats, rather than actions, become irrelevant. BDA's other side, figuring out what has happened, and the ability to continue operations are still issues. Finally, both sides should have learned by the outbreak of cyberwar that they cannot disarm cyberwarriors.

The tendency for escalation and counterescalation to descend into cyberwar may be related to the motive for the original attack. If the initial attack was somehow in error or poorly considered (the attacker underestimated the likelihood or consequences of getting caught), the escalation cycle could be self-induced. The initial attacker may gain little from going on but takes umbrage at being punished and counter-retaliates as if to say: You cannot do that to me. An attacker trying to get the target to blame someone else may respond to escalation with fake umbrage to maintain its innocence. Attackers who act out of calculation (such as by trying to ruin the target's commercial reputation, getting the target to fear networking) appear more likely to calculate that a full-blown exchange is a losing proposition. Finally, attackers with more-serious motives may have put themselves into positions from which they cannot very easily back down. Those who would disarm an opponent's military are probably ready for a fight in both the cyber and physical dimensions. The coercion motive almost begs for descent into cyberwar. A state that uses the attack to coerce and then fails to respond to retaliation reveals its weak hand, its goal utterly frustrated. That state may instead feel a great need to show that it cannot be coerced.

The Plausibility of Cyberwar

Can a scenario in which each side knowingly and even overtly carries out attacks on the other in cyberspace but refrains from physical violence be plausible? Since passing the plausibility test—that the hypothesis is worth considering—does not require surmounting high hurdles, the short answer is yes. States may plausibly attack one another in cyberspace and may be mutually unwilling to take things to the next level.

War is generally irrational, often for both parties—yet history shows that states fight one another anyway. Coercion is still a viable motive for aggression. So is the desire to assert status in international relationships and to teach lessons to other countries. It is virtually impossible to take land by cyberwar, which is fine: Land has mostly gone out of fashion as a motive for conflict. Cyberwar looks cheap: Hackers are not at personal risk. Target systems may well fail, and the effects may be highly public and consequential (albeit short-term). Overall risks may be viewed as controlled—and if initial sallies fail, their existence can be denied.

Wars, in general, are most likely to start when one or both sides seriously misestimate the results of starting a conflict. If nothing else, the narrative so far suggests the many ambiguities that attend conflict in cyberspace: doubts about the ability to discover who did what (or hid the evidence thereof), weapon effects (both prospective and retrospective), recovery time, the ability to continue similar lines of attack, cascading failures (or the lack thereof), the ability to route around damage, or the actions of third parties. If all parties react to uncertainty by shrinking from conflict, crises can be avoided—but it is not unknown for one or both sides to swallow their doubts (a common consequence of groupthink) and charge ahead.

Highly asymmetric outcomes are possible (which can then retroactively justify the venture): A hungry state may mobilize enough clever people to do serious damage to a state that is richer and more high-tech but that is correspondingly more dependent on its information systems. Similar aggression in real space would lead to crushing defeat of the smaller by the bigger. If the aggressor has few cybertargets at risk and if the richer state is disinclined to shed blood and has no other good pressure points, the latter may thereafter give the hungry aggressor wide berth and limit itself to avoiding trade or other mutually productive activities with each other. The latter is not unknown, even in today's rather peaceful world.

The enabling condition for strategic cyberwar between two states is mutual confidence that the conflict will not get physical. It is quite plausible for the target state to calculate that it is more cost-effective to bulwark its own defenses and limit its response to cyberspace than to start a shooting war, which would put serious amounts of blood and money at risk and whose ending not always easy to engineer.

The Limits of Cyberwar

There are reasons to doubt that cyberwar has what it takes to coerce a state. Casualties are the chief source of the kind of war-weariness that causes nations to sue for peace when still capable of defending themselves—but no one has yet died in a cyberattack.

The coercive effect of cyberwar has to be calculated on the basis of what one side is demanding and how badly the defender wants to avoid being known as capable of being coerced. If the stakes are high enough, a society, even a Western one, can take a great deal of punishment and still not yield. One can hardly compare what even a vigorous cyberwar might do to what the inhabitants of Sarajevo had to endure in 1992 through 1995 or to what the denizens of Jerusalem endured in 1947 and 1948. In both cases, solidarity held. Few nations have yielded to trade embargoes alone, even to universal trade embargoes. It is unclear that a cyberwar campaign would have any more effect than even a universal trade embargo, which can affect all areas of the economy and whose effects can be quite persistent.

Even a complete shutdown of all computer networks would not prevent the emergence of an economy as modern as the U.S. economy was circa 1960—and such a reversion could only be temporary, since cyberattacks rarely break things. Replace "computer networks" in the prior sentence with "publicly accessible networks" (on the thinking that computer networks under attack can isolate themselves from the outside world) and "circa 1960" becomes "circa 1995." Life in 1995 provided a fair measure of comfort to citizens of developed nations. Finally, low-tech states are inherently more immune than high-tech states and are therefore less susceptible to damage.

The notion that states can limit the damage from cyberwar through the simple (but hardly costless) expedient of air-gapping their networks suggests that the damage from cyberwar attacks may be self-limiting in ways that do not apply to other forms of coercion. Since cyberattacks require vulnerabilities to exploit, the faster and harder the attacks, the fewer good vulnerabilities remain left to exploit and the faster the potential for further loss dwindles. There may well be an effective upper limit to the cumulative damage that even the most intensive cyberwar on *core subsystems* can cause. In contrast, recurrence is more likely for attacks on *peripheral subsystems* (user machines) because the ability to take advantage of human vulnerabilities (e.g., allowing a bot to take over your computer, yielding a password to a phishing attack) appears endless. Conversely, even a full-court press against such vulnerabilities may not permit great damage. Despite the constant rise in the volume and sophistication of cyberattacks over the last 15 years, there is little evidence that they have even slowed down, much less stopped, the expansion of networks.

Another weakness of cyberwar as a coercion strategy is that the object of coercion, the state, is not the same as the most tempting targets of cyberwar: infrastructure owners and banks, most of which are privately owned and operated. In the face of cyberattack, governments could well redirect popular ire to infrastructure owners whose poor defenses allowed the public to feel the inconveniences of cyberwar. This would permit the government to maintain its (foreign) policies and duck public discontent.[6] Although infrastructure owners can refute such arguments, their refutations tend to be technical and weak.[7] Government computer systems, for their part, are not as tempting a target for cyberwarriors who seek to coerce the public. Most government computer systems can go down for several weeks with only minor inconvenience to the average citizen. The primary exceptions here may be Fedwire (because of follow-on effects on financial markets), Social Security, the Global Positioning System (heavily protected for military reasons), and the Federal Aviation Administration's air traffic control system (whose primitive nature makes it more invulnerable to attack than its rickety structure would otherwise suggest).

Can strategic cyberwar affect the outcome of, for instance, proxy wars fought in third states (such as when two powers take opposite sides in a civil war) or naval confrontations (between the United States and Iran in the Persian Gulf, perhaps)? A cyberwar that hits home might have a greater influence on the relationships between two powers than

---

[6] This assumes that voters buy that argument. Voters tend to credit governments for good times and blame them for bad times, even if the circumstances have little to do with government policy or competence. The popularity of Russia's Vladimir Putin (and, to a lesser extent, Venezuela's Hugo Chavez) was bolstered by high oil prices. In this country, the outcome of presidential elections best correlates with the rate of economic growth during the previous 12 months.

[7] One counterargument is that government-induced regulation forces infrastructure owners to facilitate interconnection with third-party providers within multiple levels of what would otherwise be a tightly integrated sector. Opening these interconnections, in turn, provides openings for mischief.

a battlefield war would, despite the casualties the latter would cause. Yet attacking the homeland, even if only in cyberspace, might elevate the importance of achieving military goals in faraway lands because the contest may be viewed as correspondingly strategic.

The Conduct of Cyberwar

The combination of inexperience and the considerable variation in motives that one state may have with which to pressure another suggests that identifying a canonical form of conflict is premature. Both sides will be making it up as they go along. Students of crisis management suggest that most world leaders are likely to proceed incrementally into crisis, assessing carefully calibrated and implemented moves, and observing and evaluating the effects.[8] Leaders tend to fear losing control of a crisis, either by engaging in behavior that may provoke unpredictable behavior from the other side or by empowering third parties to pursue their own aims irrespective of the demands of crisis management. Analyses of the Vietnam War suggest that leaders fear the loss of control more than they fear the loss of war.[9]

At first glance, cyberwar lends itself to an incremental approach because it presents such a broad range of options for contemplation. Attacks may range from introducing modest but inexplicable changes in the enemy's information systems to wholesale attempts to collapse all the adversary's infrastructures at once. At second glance, an incremental approach may be wrong. The relationship between effort and effect is highly unpredictable and may be nonlinear—perhaps nothing, followed by ratcheting up, then again nothing, followed by ratcheting up, then almost imperceptible annoyance, more ratcheting up, and nothing again, yet more ratcheting—and suddenly the players have

---

[8]   See the discussion in P. Morgan, 1977, Ch. 7, pp. 149–204.

[9]   See, for instance, Leslie Gelb, *The Irony of Vietnam: The System Worked*, Washington D.C.: Brookings Institution, 1979, and J. R. McMaster, *Dereliction of Duty: Lyndon Johnson, Robert McNamara, the Joint Chiefs of Staff, and the Lies That Led to Vietnam*, New York: Harper-Collins, 1997.

crossed some red line into strategic-level conflict. Incremental efforts may thus not produce incremental effects.

The classic case for starting a conflict with a series of probes is to use the early phases to learn how to employ one's own weapon systems, test the opponent's defenses, and discover its weak spots—feeling your way forward. Presumably, practice makes—well, if not perfect, then at least predictable. In cyberwar, however, weapon effects cannot be considered independent of the adversary's vulnerabilities and its ability to recover. Both sides are learning at the same time. For this reason, early probes, feints, and jabs may be informative only at the gross level: Is the adversary an easy or tough opponent? At the operational level, opponents may know less after these early moves then before because the terrain has changed radically in response to them. With all this, there may be no good *operational* reason not to throw everything at once—surprise has a great operational advantage here—but every good strategic reason not to. If cyberattacks are a sideshow to a shooting war or if cyberwar is deemed inevitable, strategic cyber considerations may be secondary, and operational logic in cyberspace may be sufficient. This will not be so if conflict is limited to cyberspace.

Cyberwar as a Warning Against Cyberwar

Strategic cyberwar can be used to put others on notice that their systems are not so reliable that they can afford to engage in such a fight. Consider this scenario. In step one, an attacking state flicks the lights of a big city in a target state (or induces some other obvious anomalous behavior). The act (rather than the attacker, which is kept as ambiguous as possible) gets the attention of the leaders of the target state, which perceives its infrastructure to be vulnerable.

Subsequently, that leadership, backed by the infrastructure (or major network) owners and their engineers, vows that such an act will never happen again. They get large sums of money to work hard on the problem. After this team starts to claim success, the attacker again flicks the target's lights or does something somewhere that is comparably noteworthy; this signals that vulnerabilities persist (admittedly, step two is hard, precisely because the target state is working diligently against the possibility). This not only reduces the credibility of the target's information system security, it also, and more importantly, reduces the credibility of those who promised to achieve that security.

Yet the attacker does not reveal itself. This is unnecessary and even gets in the way. Doing so would make getting back at the attacker a more-visible centerpiece of the target's strategy than reassuring the public, albeit misleadingly at first. Indeed, the attack itself is not so much the issue as it is to foster a general sense that the other side's information systems are fragile and unreliable.[10] The attacker's message then becomes not "Cower before us!"—which requires identifying "us"—but the more impersonal, "You live in glass houses; are you sure you want to invest so much in stones?"

Preserving a Second-Strike Capability

As noted in Chapter Four, one of the greatest challenges in carrying out a cyberattack campaign is ensuring that the first strike does not create conditions that blunt the effects of a second strike. Certainly, that is challenging enough by itself.

Attackers can, however, take steps to retard victims' efforts to make themselves less vulnerable, such as the following:

- Induce errors that look as though they could arise from software failures and transient conditions, rather than from attacks as such.
- Find ways to probe the targeted system for a reaction that is less likely to induce changes in the system as a response (defenders are less likely to make radical changes if they believe they have defeated such probes).
- Attack system-specific vulnerabilities rather than generic vulnerabilities (the latter, when patched, make many systems harder to attack again).

---

[10] Cineastes may recall the last few minutes of the Gene Hackman movie *The Conversation*, 1974.

- Seek out who are unlikely to share their experiences with others; this way, a given exploit may be used again on another target.
- Use exploits that are likely to become obsolete and hence useless soonest (those with later use-by dates can be saved for later contingencies).
- Take advantage of the sort of vulnerabilities that only a painstaking search can uncover. This is one advantage of a supply-chain attack on software or hardware: It forces a thorough review of many components. Code that has been implanted months or years previous may have similar advantages.
- Find attacks that are relatively insensitive to simple countermeasures, such as disconnecting systems that really should not have been connected in the first place.

All this is easier said than done, by the way.

Sub-Rosa Cyberwar?

Coercion—especially against democratic states—normally requires the damage to be publicly visible and clearly associated with the coercer and its cause. Adversary actions need not affect the public if there are other ways to compel governments to accede to demands. Indeed, the opposite may be true: The less the public knows, the easier it may be to garner concessions, especially invisible ones.

Cyberwar is unique in that the public need not know it is taking place—may not know what the problem is or, indeed, whether there is any problem at all. Factors other than cyberwar (e.g., error, accident) can be adduced to explain visible disruption—up to a point. Thus, a sub-rosa cyberwar is not impossible. But would it be worthwhile?

The case for sub-rosa cyberwar rests on the belief that going public would put more pressure on the state to stand firm than to concede. The attacker counts on the possibility that the target's leaders are less afraid to make concessions whose true rationale can be hidden than they are afraid to be blamed—and have no one to shift blame to— when, say, the economy hits an air pocket. As long as the new policy (which contains concessions) does not appear unwise per se or does not contradict earlier policies too much, the target merely needs to hide the fact that its policy choices were driven by fear. Keeping mum has other advantages for the target. Reducing the public itch for revenge (or its desire to demonstrate resolve) may facilitate negotiations or mutual de-escalation. Obscuring the fact or at least the damage from the attack may also mask the state's vulnerabilities from the eyes of third parties (presumably, the attacker will have a better sense of which vulnerabilities it had, in fact, exploited).

Conversely, screaming helps mobilize the citizenry to support the government and (less cynically) pay attention to information security. It raises the seriousness level of the whole cyberwar contest and thus gives the government more scope for implementing domestic security measures that the citizenry would otherwise object to. If the fact of the damage is evident, but not the cause, revealing the cause may enhance the credibility of infrastructure owners by switching attention from their own fecklessness as sysadmins to factors (portrayed as) outside their control.

A Government Role in Defending Against Cyberwar

Apart from protecting its own systems, the most obvious ways that government can defend against cyberwar are indirect: Sponsor research, development, and standard creation in computer network defense. Maximize the incentives for private industry to keep its own house in order. Increase the resources devoted to cyber forensics, including the distribution of honeypots to trap rogue code for analysis. Encourage information-sharing among both private and public network operators. Invest in threat intelligence. Subsidize the education of computer security professionals. All are current agenda items. In a cyberwar, all would receive greater emphasis.

Are there any *direct* methods the government can use to protect the state? Inasmuch as cyberattacks require vulnerabilities to exploit—and vulnerabilities, if they exist—are faults of the target systems, the fundamental answer has to be *no*. Operators must see to the vulnerabilities of their own systems.

This tenet admits of two possible exceptions: combating DDOS attacks (which, uniquely, do not arise from target system faults) and mounting a broad-scale malware defense. The two are related; over the last few years, hackers have embraced a client-side attack technique that puts rogue code into email attachments and Web pages. Users who download the content unwittingly introduce executable computer code into the their machines, which later attaches itself to the boot partition, the operating code, or a common application (e.g., Internet Explorer). Running the computer or opening the application also runs the payload code. Some code turns the system into a bot. Other code exfiltrates information (Web-surfing habits, in the case of spyware). Such code can also be written to carry out cyberattacks.

Because the set of vulnerabilities is limited, the set of exploits that take advantage of them is also limited. The active components of such exploits have signatures in the same way that a virus (be it carbon- or silicon-based) does. Thus, exploits can be detected and neutralized using similar techniques—for example, detecting documents that contain suspicious byte patterns and neutralizing such patterns (e.g., removing them or converting them into benign form).

Why engineer a national capability to detect bugs when so much private capability already exists? First, no network security company or network operator has the quality and quantities of resources available to a national government—here, economies of scale are important. Second, the processing power required to analyze documents against a vast variety of hazards may exceed what individual machines or even individual routers can handle comfortably—here, too, economies of scale matter.[11] Third, not everyone buys protection, but those who refuse to do so imperil others—hence the case for government intervention. Fourth, the resistance of a state to cyberwar is a function of the resistance of each of its institutions; the whole is greater than the sum of its parts.

Thus, there is something to be said for a wide-area filtering capability for packets that come from outside into the state, if the cybersecurity

---

[11] Up to a point—driving a state's traffic through one gateway to realize economies of scale for analyzing packets creates a single point of failure.

problem so warrants and as long as entry code plays a significant role in how systems are breached. But is the state the right place for the filter? First, only internal scrubbing can catch internally resident malware (external threats can evade filtering as long as one internal system translates encrypted and thus unfiltered malware into a document with rogue code). Second, a great deal of encrypted traffic crosses national borders, notably within intracorporate virtual private networks, and Secure Hypertext Transfer Protocol traffic (used, for example, for protected e-commerce). Furthermore, whoever analyzes traffic for rogue code also has the capability to capture intended content for any variety of purposes (the nature of which can be left to the imagination).

A somewhat less-problematic approach is to apply techniques similar to those for a coalition of the willing, creating a cyber enclave including the U.S. government, friendly governments, and key infrastructure providers for such states—precisely the targets of what might be a cyberwar. Careful delineation of the enclave's "borders" should reduce the amount of encrypted material that has to be dealt with, and defining the enclave in terms of institutions minimizes having to cross virtual private networks (encrypted extranets that link buyers and suppliers are still a problem). An enclave of institutions will not stop DDOS attacks (most bots lie outside such institutions) but can permit communications within the enclave while such attacks are going on. Within such institutions, issues of trust (reading the packets) can be dealt with.

Managing the Effects of Cyberwar

The course of cyberwar depends on the systems being targeted, which in turn will depend on which ones have what vulnerabilities and to what extent exploiting them can discomfort the state. Cascading or ripple effects are often a bonus. Good attack methods work in nonobvious ways, have few good countermeasures, do not exhaust vulnerabilities that may come in handy for operational cyberwar, and can thus be used repeatedly. Attacks on system peripheries are less powerful than attacks on system cores but can be repeated more often.

Although nuclear warfighting pays attention to keeping some targets in reserve (lest the other side, having nothing left to lose, has no reason to yield or quit), there is no easy analogy in cyberspace. One can whack every conceivable cybertarget one week, and all could be back in service the next week, available to be whacked again (albeit with more difficulty[12]). Even never-struck targets are likely to be better defended—the flare-up of cyberwar would be expected to persuade all system owners to pay more attention to security.

Some targets may be too risky or messy to be good targets. The risky targets include nuclear command-and-control systems (lest nervous adversaries conclude that they must use it or lose it) and space systems (many of which are also strategic). Targets that give pause because of the mess their confusion may cause include those whose malfunctioning may lead to civilian deaths, those whose disruption can create vast environmental damage, and those whose integrity and accuracy can be very difficult to restore when peace resumes (e.g., managers of bank and billing records). It would be good to think that such systems are unassailable (or at least engineered to fail safely) precisely because they are sensitive.

Might a better reason to leave targets untouched be that restraint might persuade the other side to do likewise, thereby limiting mutual destruction?[13] Mutually respected safe zones may even provide a path for both sides to de-escalate.

Unfortunately, communicating such reserve in cyberwar is likely to be harder than doing so in physical war, where hits are obvious. A target untouched by hacking may communicate reserve on the attacker's part—or that the target is too hard to hit. Conversely, failed attempts, even if detected, do not prove that the adversary was attempting to violate such a zone. As noted in Chapter Four, an attack may overfulfill its

---

[12] Conversely, implants—while they remain undiscovered—make it easier to access systems the next time.

[13] From Schelling, 1966, p. 191: "We usually think of having failed if a major war ever occurs. And so it has; but it could fail worse if no effort were made to extend deterrence into war itself."

quota, so to speak, and interfere with systems that were not supposed to feel even indirect effects from that attack.

Third parties may complicate achieving finesse. Even with the optimistic belief that states alone have the capability to strike certain targets, the list of those who can make *failed* attempts to strike certain targets may include third parties. Indeed, third parties in general can complicate signaling in cyberwar just as they can with cyberattacks in peacetime. It would benefit both parties in such a contest to make special efforts to distinguish each other's work from that of hackers. Such differentiation would help gauge the strategy and relative efficiency of adversary attacks, assign blame correctly when something is hit, sense possible adversary de-escalation, or discount what may otherwise seem to be adversary escalation. Although there may be times when overt cybercombatants would prefer that their attacks appear to come from somewhere else, each may want to inflate its strength and thereby get credit for attacks it did not pull off.

Any unilateral effort to control escalation by threatening retaliation if specified lines are crossed faces obstacles similar to those that bedevil attempts to deter cyberattacks in the first place. True, attribution and establishing the will to retaliate would likely pose less of a problem once the two admitted that they constantly hack each other. Gaining consensus on thresholds, however, may not be possible, even between two rational and reasonable combatants—a quality not always present between those locked in combat. Unilateral declarations on what is and is not off limits may not work if the other side believes that the distinctions drawn favor the first side;[14] it gets worse if the first side has *already* attacked targets the other side believes are of comparable sensitivity to those the first side would place off limits. If the thresholds are expressed in terms, not of categories, but of effects (e.g., the presence or, even less clearly, the extent of casualties), the problem of intentionality rises—"How was I to know that attacking your systems

---

[14] The first side may declare any cyberattack that may pose a risk of catastrophe off limits. It might, say, put dams in that category. The other side notices that the first side gets most of its electricity from hydropower, while it gets most of its own electricity from coal-fired power plants that do not merit such protection.

would put lives at risk? Attacking ours would not have put lives at risk." The same holds for BDA: If a victim is to scream, "Foul!" it must demonstrate not only which systems have been hit but how they were damaged—a sweet hint for the attackers, always unsure of what they have done. Further complications arise for cyberwar carried out against a background of bloody hostilities elsewhere; how does one establish comparability between real and virtual effects? Someone contemplating cyberwar can therefore entertain no reasonable hope that it can be controlled at some level short of a full-court press by both sides. Ambiguity does not leave room for nuance.

Ironically, shaping a cyberwar to permit the other side to recover quickly is easier than shaping kinetic conflict. Promising that recovery could be rapid under the right circumstances might induce foes to settle. True, this consideration rarely moderates the conduct of physical warfare (even when the winner strongly suspects that it will have to pay for cleanup, such as after Operation Iraqi Freedom), but this may reflect the close relationship between the damage an attack caused and the cost of repairing it once peace breaks out.[15] Rejecting certain candidate targets because they would be hard to repair means reducing the pressure on the other side to settle.

Cyberwar, however, provides unique opportunities to use such promises constructively. This is because the art of a cyberattack lies not in destroying but confusing target systems. Systems can be set straight painlessly, by revealing the sleeper code that was placed, which algorithms were corrupted and how they can be restored, or what key was used to encrypt and make inaccessible which data files. Promising such "repairs" is inadvisable, however, if renewed hostilities are likely soon.[16]

---

[15] For instance, a physical attack that caused environmental damage may do little to disarm the enemy but be quite expensive to clean up completely. Saddam Hussein's setting Kuwait's oil fields on fire comes close, here, but that was clearly seen as barbarous in the West.

[16] The concept "soon" can be measured in terms of software replacement cycles (typically measured in months or years), on the theory that tricks depreciate and have to be renewed periodically anyway. Thus, whether or not one revealed the vulnerabilities and exploits present in today's system may not matter much ten years hence. By then, it is likely that most

Terminating Cyberwar

War strategies are ultimately about war termination. Cyberwar, as noted, is highly unlikely to be terminated because the adversary has been disarmed (much less overturned) by force. Such wars are more likely to end by exhaustion or by concessions. Unfortunately, the longer wars go on, the less they are about their original aims and the more they are about themselves (e.g., revenge and, less irrationally, the mutual desire of each side to ensure that it is secure from the other). Cyberwar presents an additional and compelling challenge: How can one tell that the other side has, in fact, stopped its attacks?

Consider three war termination paths: negotiation leading to termination, tacit de-escalation, or petering out. Monitoring peace pacts in cyberspace poses challenges not found in physical space. If either side still believes it can, if unpunished, reap unilateral advantages from an attack, attribution and BDA will likely remain as difficult afterward as they were beforehand.[17] Each could cheat by shifting from visible disruption attacks to more-subtle corruption attacks. Both sides also must contend with the possibility of potential mischief from third-party (including state) hackers masquerading as the other side.

In physical wars, peace pacts are often followed by unilateral disarmament (after World War I, for instance, Germany's army was limited to 100,000 soldiers) and multilateral disarmament (for example, the Washington Naval Treaty). But disarmament in cyberspace is virtually meaningless because cyberwar is less about arms (exploits) than about vulnerabilities. Appendix C discusses arms control in more detail, but the conclusion remains: Disarmament cannot bulwark a peace agreement.

Mutual transparency may help keep the peace (in much the same way that formerly warring sides exchanged hostages), but no state (not even a friendly one) exposes the secrets of its security architecture to

---

old vulnerabilities would have been fixed only to be replaced by new ones, and new exploits would have to be discovered.

[17] If *attack* is defined here to include CNE, the odds that one side may find further attacks useful only go up.

another. If it did, the transparency would have to be bilateral rather than public, lest mischievous third parties profit from the newfound knowledge. Even then, each side could attack the other from third parties outside the transparency agreement. Thus, the best outcome is for neither side to find any especial reason to commit serious resources to breaking the systems of the other. This may ensue because the broader ends that led at least one of them into cyberwar in the first place have been met or because further cyberwar will get no party closer to meeting them than the last spate of cyberwar did.

Mutual tacit de-escalation, compared with explicit war termination, has the advantage of not requiring formal adjudication of the original issues. Although the original issues may remain unresolved, the prospects for peace may rest on both sides believing that neither would make much headway through further cyberwar. Unfortunately, tacit de-escalation presents many of the same validation problems as negotiations—only made worse by the fact that there would only be a rough consensus rather than an explicit statement of what (e.g., which targets) was and was not considered a violation. How could one tell that the other side is even cooperating, absent clarity on what constituted cooperation?

The third path to peace, or at least a return to the muffled din that preceded the outbreak of cyberwar, would be for attacks on one or both sides to peter out. Each side could unilaterally conclude that cyberattacks were growing more difficult to conduct, showing decreasing returns for the effort, and thus pointlessly irritating the opponent when greater cooperation might be called for. The part of the equation in which one side decides that the effort no longer pays is not strategically problematic because it does not require the other side to recognize that the first side has changed its behavior. But it is hard to believe that the party that quit making the effort would not hope to see some rewards for its restraint. As long as the one side had made neither explicit (that is, negotiated) nor implicit commitments to restraint, the other side would not be able to hold up some future system malfunction as evidence that it had been lied to or cheated. Furthermore, if the other side still found advantage in computer attacks—or if it was engaged in other forms of hostilities—it may have no motive to acknowledge such restraint. But if the other side also finds that the advantages of hacking have waned or that they are trumped by the rewards of friendly engagement, it too might work itself into a modus vivendi.

Conclusions

Cyberwar is likely to be problematic largely because the coercive effects of cyberattacks are speculative. As a threat, it may not be believed; as a reality, it may not cause enough cumulative damage to make the target cry uncle. Any technique is likely to simultaneously frighten and anger its victims. The balance between anger and fright has everything to do with whether the first attack is a harbinger of worse or inherently not repeatable to the same degree—a primary characteristic of strategic cyberwar. With strategic cyberwar, the fright vector is likely to be weak while the anger vector is unlikely to diminish—a poor calculus from the attacker's perspective. Although these may not be reasons to abstain from strategic computer network attack, they do suggest that cyberwar is neither a good adjunct to nor an adequate substitute for more-conventional forms of strategic coercion. It is thus hard to argue that the ability to wage strategic cyberwar should be a priority area for U.S. investment and, by extension, for U.S. Air Force investment. It is not even clear whether there should be an intelligence effort of the intensity required to enable strategic cyberwar.

CHAPTER SEVEN

Operational Cyberwar

Operational cyberwar consists of wartime cyberattacks against military targets and military-related civilian targets.[1] Even if this does not constitute raw power, it can be a decisive force multiplier if employed carefully, discriminately, and at precisely the right time.

For this discussion, the context is a conflict between the United States and an opponent that also relies heavily on computer networks to conduct military operations (a growing candidate list as digitization becomes the norm). Because operational cyberwar against military targets is not an escalation of physical warfare, it does not raise broader questions about the depth of the war. Those who accept the destruction of information systems as legitimate can hardly quibble about operations to confuse such systems. Thus, this discussion can avoid delving into strategic issues (until the matter of attacks on military-related civilian targets is raised). Operational cyberwar is also not the same as CNE, even if CNE is required to understand the target, get access to the right attack vantage point, and collect BDA. To keep the lines of argument clear, the discussion also excludes (1) physical attacks on networks (excepting rare cases when a cyberattack leads to physical

---

[1] While this does introduce the questions that make cyberdeterrence problematic, these issues would have little effect on the tractability of operational cyberwar. The difficulties of conducting BDA and continuing attacks might retard operational cyberwar but not stop it. Physical combat makes attribution and threshold issues irrelevant. The fortification (moral hazard) issue relates solely to private infrastructures. The fact of physical warfare means that matters have already escalated beyond cyberwar. Emerging warriors can still blur signals, but signals matter far less once hostilities have started. Finally, incomparability of cyberspace targets is quite secondary if the war is being decided by kinetic combat.

damage), (2) electronic interference against a network's RF links, and (3) psychological operations (even if cyberoperations do have psychological effects).[2]

Stipulating that the opponent has networks is necessary to give meaning to operational cyberwar. Indeed, that is what differentiates this from other forms of combat. U.S. air and space capabilities are, if anything, more dominant against adversaries lacking aircraft or spacecraft than they are against adversaries with them. But the contribution that U.S. dominance in operational cyberwar can make to victory reflects the extent to which adversaries have a footprint in that domain: no footprint—no impact.[3]

Beginning this discussion requires clearing the air on two matters. First, operational cyberwar cannot win an overall war on its own; it is a support function, and is likely to remain so indefinitely. It cannot occupy territory; put people's lives at risk; or, except in specialized cases, break things.[4] The direct effects of the most fiendish cyberattacks, *if*

---

[2] Again, the required distinctions may be quite subtle. An unsigned mass email message sent to opposing warfighters calling on them to lay down their arms would be a psychological operation. An email message that calls on them to lay down their arms that appears to come from their commanding officer would be an act of deception, which would better be considered part of operational cyberwar. The first uses clever content to work its magic; the second uses hacking techniques to give its message the patina of authority.

[3] Now that the global population of cell phone owners has exceeded the global population of those without cell phones, the odds of finding a military force without access to cyberspace may well be approaching zero. That being so, the relevant measure is the degree of dependence on, rather than familiarity with, cyberspace. (MOCOM2020 Team, "4.1 Billion Mobile Phone Subscribers Worldwide," MOCOM2020 Web site, March 27, 2009; International Telecommunications Union, "New ITU ICT Development Index Compares 154 Countries: Northern Europe Tops ICT Developments," press release, Geneva, March 2, 2009.)

[4] It can, however, corrupt systems so much that restoring them may cost more money than it took to build them in the first place. These days, a system that has been attacked and compromised can be restored within hours or days, if it is mission critical. If it is not mission critical (e.g., office automation systems for education and administration) but handles sensitive information, administrators may prefer to disable it for months until they are certain that sensitive information cannot be compromised again (following a cyber attack, the National Defense University's network was reportedly taken down for an extended period to replace its hardware; see Bill Gertz and Rowan Scarborough, "Inside the Ring: NDU Hacked," *Washington Times*, January 12, 2007). Given the years it takes to build, test, and accept a

discovered, can often be reversed within hours or, at most, weeks.[5] As previously noted, cyberattacks are likely to be too weak to coerce a population into surrender, particularly one already hardened by the normal privations of war. A support function is hardly a euphemism for a worthless endeavor, though. The current U.S. space constellation is a support function but is also indispensable to how the nation would wage conventional war. The Middle East has provided many examples of how airpower can convert the prospects of slow heavy combat into a rout (the 1967 Six-Day War, the 1991 Gulf War). But this does mean that operational cyberwar can be analyzed only in the context of the military functions it *does*, in fact, support.

Second, the question of cybersupremacy is meaningless and, as such, is not a proper goal for operational cyberwarriors. Here, we define supremacy as being analogous to its use in other media[6]: One air force can prevent another from taking to or at least surviving in the air; one navy can bottle up another in port; one army can prevent another from holding ground.

Cybersupremacy is impossible because cyberspace is not a unitary domain. Both organizations can simultaneously keep each other off their own networks. In practice, hackers do get into other people's networks. Unfortunately, the idea that someone "owns" another network if he or she can make its machines obey his or her instructions abuses the concept of ownership. Ownership implies exclusivity. If nothing else, outside hackers cannot claim physical control, and physical con-

---

really complex information system, it is easy to see how a cyberattack could expose such fundamental weaknesses that it would take years, rather than months, before it can resume service at its previous level of trust. If there were little left of the original system to recover, a cyberattack might, in effect, destroy a computer system. That noted, real-world examples of such phenomena are scarce.

[5] In the unlikely but not impossible event that a computer network attack caused equipment to break, the effects might take much more time to reverse, especially if repairing it or finding parts for it is difficult.

[6] In theory, *space supremacy* would mean that one state's space constellation could knock another state's spacecraft from orbit. However, terrestrial weapons can also knock spacecraft from orbit (the first U.S. spacecraft shot down was hit by a missile launched from an F-15; the second, in 2008, was hit by a ship-launched missile).

trol dominates all other forms of control. Owners can physically add or remove machines from a network and can install software directly. If worse comes to worst, owners can discard and replace systems. Owners with the wit to have backed up their data and applications can resynthesize their networks regardless of who has messed with them. Indeed, a large percentage of exploits require physical access to a system to work.[7] Furthermore, there is no ipso facto relationship between keeping the bad guys out and getting into where the bad guys live—even if such underlying factors as the relative quantity and quality of each others' hackers predisposes success or failure at both. In short, there is no such thing as a single cyberspace, but at least two: yours and theirs. Without a *common* space, there is no such thing as supremacy.[8]

The remainder of the chapter discusses some of the operational challenges of operational cyberwar. Cyberwar can play three key roles: It might cripple adversary capabilities quickly, if the adversary is caught by surprise. It can be used as a rapier in limited situations, thereby affording a temporary but potentially decisive military advantage. It can also inhibit the adversary from using its systems confidently. Following discussion of these roles, the chapter touches on civilian targets before taking up organizing for operational cyberwar.

---

[7]   Ed Felten demonstrated a way of capturing the memory from a machine, but this method requires dropping the memory into dry ice within seconds of its having been turned off. See J. Alex Halderman, Seth D. Schoen, Nadia Heninger, William Clarkson, William Paul, Joseph A. Calandrino, Ariel J. Feldman, Jacob Appelbaum, and Edward W. Felten, "Lest We Remember: Cold Boot Attacks on Encryption Keys," *Proceedings of the 17th USENIX Security Symposium*, Berkeley, Calif.: USENIX Association, 2008, pp. 45–60.

[8]   Can one define cybersuperiority over the part of cyberspace that neither side controls? This is a largely notional issue, in the sense that very few militaries make use of the cyberspace commons to do their serious work (although there is a global Domain Name Service, DoD owns the ".mil" domain). Furthermore, every network belongs to someone already. Thus, although one may contemplate two sides competing to control third-party networks, these third parties might actually have a word or two to say about the matter—and they have physical control and authentication in real space on their side.

Cyberwar as a Bolt from the Blue

Cyberattacks are about deception, and the essence of deception is the difference between what you expect and what you get: surprise. This is why operational cyberwar is tailor-made for surprise attack and a poor choice for repeated attacks: It is difficult to surprise the same sysadmin twice in the same way.

A surprise cyberattack might remove capabilities that the adversary relied on to complete its military missions. Such capabilities can be defensive—perhaps a SAM failed to engage its target correctly or at all. They can also be offensive—perhaps the command and control required to synchronize an invasion fleet is suddenly crippled by a consequent loss of efficiency, hampering synchronization and coordination. Success requires only that the adversary be surprised by an exploit targeting a vulnerability it did not realize it had or did not think anyone could utilize. However, the effects of any one attack may be enhanced if the *possibility* of an attack is also a surprise.

Surprise, at both the strategic and operational levels, works differently in cyberspace than it does in physical space. Start with strategic surprise as an analog to the use of cyberwar just prior to or at the outset of a military engagement. As Richard Betts has concluded, "[t]here are no significant cases of bolts from the blue in the 20th century. All major sudden attacks occurred in situations of prolonged tension during which the victim state's leaders recognized that war might be on the horizon."[9] Correspondingly, a cyber bolt from the blue originating in a state that did not have tensions with the target would be historically novel.

Several plausible surprises are nevertheless possible. First, an attacker could launch a military confrontation during a period of tension by attacking the civilian infrastructure. Here, the surprise would be less the fact of war than the means of pursuing it—an immediate shift to the strategic level. Many military surprises appear in retrospect to have succeeded because attackers found unexpected ways to neutralize disadvantages that the victim thought should have precluded

---

[9]   Richard Betts, *Surprise Attack*, Washington D.C.: Brookings Institution, 1982, p. 18.

action.[10] In this case, it is difficult to think of how a cyberattack on civilian infrastructure would reduce the victim's military efficacy or its top-down command and control (unless military operations could not be carried out if civilian telecommunications were down). Starting at the strategic level also threatens strategic retaliation from the outset (possibly trumping on-the-ground gains). This surprise, then, appears to be irrational. However, as Richard Betts has observed, "[a]pparently irrational behavior is one of the most important elements in several past surprise attacks."[11]

Alternatively, and more plausibly, a bolt-from-the-blue cyberattack could be launched just prior to or simultaneously with a surprise military attack. Although the military may be unprepared for war (political leaders may have refused it permission to deploy forward aggressively or to mobilize sufficiently), it has no one but itself to blame for its being unready to support fielded forces with information systems. Even if the resources available for cybersecurity for the long haul are deficient for political reasons, much of what militaries can do to minimize the damage from a cyberattack can be done in days or weeks and with few resources. Militaries can install patches, run tests, map their own networks (e.g., to find unnecessary portals or unsecured machines), pursue anomalies aggressively, restrict unnecessary access privileges, etc. Similarly, any self-respecting military should expect to be the target of state-sponsored CNE at all times. It is constantly probed by all comers—so it would, or should, be prepared to at least some degree.

In the highly unlikely event that the onset of hostilities did not proceed from the exacerbation of a crisis, even a target conscious of the threat may have peacetime and resource-constrained modes. In these, it may accept the risks of more-open access to gain its benefits (e.g., faster learning) and relax its constant testing for security vulnerabilities or corrupted files. Undiscovered or, worse, silently exploited vulnera-

---

[10] Betts, 1982, p. 14, argues that, in 1973, Egypt was not constrained by the premise that it would not attack until the balance of airpower began to turn in its favor. Instead, it resorted "to reliance on air *defense* to neutralize Israeli air power, rather than to meeting it on its own terms."

[11] Betts, 1982, p. 118.

bilities may lurk. Furthermore, a state at peace may have multiple challengers, each with different goals and a different MO. It may spread its own collection efforts against many potential adversaries. War brings focus and concentration on the one actual foe among many potential ones.

Operational surprise is still possible after war starts and presumably after the target has hardened its system against attack. The target may have yet to see exploits that the attacker has in waiting or has implanted but not yet activated. For this, the attacker needs a good bag of tricks that it has yet to use (or at least has not used widely enough for them to be recognized) against vulnerabilities the target did not realize it had (or at least has not fixed). Thus, a cyberattack can still be a bolt from the blue, even if hostilities are imminent or under way.

Because operational cyberwar can still work if there is *operational* surprise, it can still be used by states, such as the United States, whose war policies do not include *strategic* surprise—but it is even more useful for states that find strategic surprise appealing.

A technique that requires surprise to work can still be a useful in wartime—if the effects in cyberspace are rapidly and tightly integrated with effects in real space. Consider three categories of effects: eruption, disruption, and corruption.

Eruption is the temporary but intense virtual illumination that operational cyberwar can bring to the battlespace to highlight the presence and location of military targets for immediate destruction. Making adversary systems light up may also help in discovering "hiding places" which thereafter cannot be used again, or at least not so easily.[12] If the cyberattack works spectacularly well, it may be possible to estimate adversary strength by counting what lights up.[13] One might also get a sense of the enemy's immediate strategy from knowing how its forces are arrayed. Eruption, though, depends on two assumptions. One is

---

[12] This assumes the "hiding places" are not so buried as to mask any signals emanating therefrom.

[13] Doing so confidently requires either that nearly everything lights up or that what lights up is truly random and that subsequent discovery provides ways to estimate what percentage of assets in a given category lit up.

that the target systems have ways to emit signals on command and they accept commands to squawk. Two is that the attacker can take such a squawk, identify its source, localize it, and get weapons on it before it runs and hides—which it will be in a hurry to do if and when it realizes that it has just given itself away. Timing is critical here. If multiple targets light up simultaneously (e.g., within a few seconds of each other), it may be difficult to distinguish one target from another—a sudden wave of electronic noise may not be terribly helpful to targeteers. There may also be limits on how many targets can be struck in, at most, the few hours that constitute a window of opportunity. Even under the best conditions, prosecuting such signals would require detailed coordination with operational units: to acquire many signals, determine which are spurious, correlate them with targets, evaluate the targets, and sequence them for prosecution. Once the target realizes that its systems are giving its position away, it is likely to shut down its communications until it has figured out why its own equipment is misbehaving.

Disruption renders military systems temporarily incapable to a greater or lesser degree, leaving a different window of opportunity to be exploited vigorously. The varieties of disruption are legion: communications that are squelched because they have squawked unexpectedly, command-and-control systems that suddenly refuse to function, sensors that go black, weapons whose electronics hang up (which prevents modern weapons from functioning, even in a debased or manual mode). Determining whether and how much disruption has taken place would be a potential challenge—"potential" in that one could just go ahead and conduct operations and pocket any enemy paralysis as a bonus. Dark lights and dead signals are a good clue. Reduced electronic traffic is almost as good. Good in-the-net intelligence or echoes elsewhere (internal complaints, for instance) are often reliable. It gets tricky if one has to determine paralysis by examining how tight adversary execution is, and trickier yet if one has to look to operational cyberwar to reduce the quality of the adversary's decisionmaking. Worse, if the target expects some sort of attack (even if does not know how or where), it may be prepared to generate spurious indications, such as that ostensibly paralyzed units are operable or that inoperable units do not appear paralyzed. Military strategies that would exploit enemy paralysis may require operators be confident that paralyzed units are, in fact, paralyzed—and that information may not be forthcoming the first time a cyberattack is carried out.

Corruption has the virtue of persisting longer the more subtle (and, thus, often the less advantageous) its effect. Some forms of corruption are operational—a missile that fails to point in the right direction; a sensor that fails to pick up on certain types of signals, seems less sensitive that it should be, or misinterprets what it sees; a communication system that misroutes packets or leaves some nodes mysteriously in the dark; a logistics system that fails to update itself consistently. Intermittent disruption may be as hard to detect as corruption if the adversary cannot distinguish a system with induced error rates from one that happens to operate at one end of its expected error band. Corruption may be harder to detect than disruption: There may be a race to see whether the attacker or target finds it first. The attacker has the advantage of knowing what kind of corruption to look for and thus has a better sense of whether sought-for errors are natural or induced. The target, however, is likely to be more familiar with the system's normal parameters, is almost always better placed to measure its performance (especially when fed test inputs), and can more easily probe the system's software. The target can also fool the attacker by creating spurious indications of corruption. Furthermore, if multiple systems are corrupted in the same way, the fact of corruption may be quite obvious (even if the source is a mystery). With all that, it may be too much to expect attackers to exploit the possibility of such corruption—faults in enemy systems can only be appreciated, not appropriated, in advance.

Simultaneous cyberattacks (parallel warfare in cyberspace) have their attraction. When many things go wrong unexpectedly at once, great confusion is likely. A foe bent on executing a plan suddenly has to confront the difficult problems of discovery, diagnosis, and triage among multiple systems that fail for no good reason. If the forensics, recovery, and mitigation resources available to address simultaneous problems are limited, the affected systems may be disabled longer than if problems had occurred sequentially.[14] If systems are cross-linked (e.g., restoring phones depends on restoring electric power, and vice versa), restoration may lag all the more. Serial attacks permit healthy systems to cover for unhealthy ones. This cannot happen if everything is affected simultaneously. Although a well-prepared or at least mentally agile commander may cope, not all commanders are so cool when facing disasters in systems they barely understand; confusion breeds paralysis, and paralysis breeds defeat.

Several disadvantages of parallel attacks have been mentioned, two of which are (1) that taking advantage of multiple failures may tax the physical resources of attackers and (2) that using any one attack method against second-order targets now may make it less useful against first-order targets that may emerge later. The advantages of parallel attacks are not so pronounced if the attacks are designed to corrupt rather than disrupt. Indeed, if effects are sought through anomalies too subtle to be recognized as being unusual, simultaneity may defeat the purpose—a correlated set of otherwise unremarkable disappointments may actually strike someone as out of the ordinary.

Although the cyber bolt from the blue need not be a sneak attack, it is easier to set up the conditions in peacetime. Even if the attack is novel, as most successful cyberattacks are, the techniques required to prepare the environment or maneuver the attack vector (e.g., the implant) into place will resemble those used for CNE. Such techniques work more easily before the target's security posture is tightened.

Putting the attack in place merits several precautions. First is to avoid creating a large spike in suspicious activity within the target network—in general, that means avoiding any spike in suspicious activity anywhere in the target network being monitored or anything else that is likely to prompt the target to check its system for corruption. Second is to carefully coordinate CNE activities, not least when using the same penetration techniques that intelligence agencies employ to gain access for collection. The risk is that discovery and elucidation of

---

[14] This refers to specialized resources that can be deployed to emergencies. As a rule, sysadmins are the first line of defense and recovery, and each system has its own administration.

the attack preparations may hint at intelligence collection penetration methods. Third is to avoid using too many implants that are so similar that detection of one will suggest what to look for in detecting others.

Once an attack is in place, though, will it work when needed? The short answer is that, for all the reasons previously discussed (see Chapter Three, "Can We Hold Their Assets at Risk?"), this cannot be determined reliably.[15] That is especially true if the definition of "work" is that the attack not only affects target systems but affects them in ways that disable the adversary's ability to wage war. Trying to ascertain that the system contains no compensating code to mask the effects of the vulnerability (or even within the module that contains the vulnerability) is essentially trying to prove a negative.

Once a surprise attack has begun, operational cyberwar is likely to change from being treated as a general-purpose weapon to being one husbanded for special occasions against special targets. Such targets would be thoroughly researched to discover uncommon vulnerabilities whose exploitation depended on crucial timing or that could be set up via social engineering or the cultivation of insiders. Potential attackers would have to know the target and its defenders, perhaps better than its intended users would. Because the scope of subsequent operations would be limited (largely to forestall depletion), they would have to be thoroughly prepared, precise, and closely monitored, since effects can rarely be assumed in advance. If operational cyberwar works, consequences almost always have to follow on its heels.

Dampening the Ardor for Network-Centric Operations

Operational cyberwar can also be used to make a foe wary of networking, in the hope that it will withdraw from the outside world, the better

---

[15] This applies to attacks rather than to CNE. The latter can be known to succeed when information comes flowing back. Determining whether the information is authentic or deceptive is another matter.

to reassure itself of its organic capabilities.[16] This is a plausible enough motive for a peacetime cyberattack, but the effect could be sharper in wartime. War promotes self-reliance, paranoia (someone *is* out to get you), and the realization that security mistakes may have deadly consequences. Organizations have a natural tendency to close in on themselves when under threat. The role of operational cyberwar is to play on such tendencies. The adversary should be made continually aware of the possibility—and especially the consequences—of a successful cyberattack. Actual attacks are less important, except as reminders. Indeed, carrying out too many attacks can backfire if diminishing returns set in, causing the adversary to conclude that the attacker is running out of tricks.

Although some tightening up is advisable as war changes the threat environment in cyberspace, the point is to make the adversary overdo things. In a cyber lockdown, more activity would be isolated within compartments, and access to them would tighten.[17] Users would find it more onerous to deal with information systems as it became more difficult to get on machines and stay logged in. Users would have less discretion to adapt systems to their needs or even to use the available tools in unanticipated ways. Fewer people would be able to access the network.

Lockdowns can wrench and perhaps break coalitions between militaries on one hand and support infrastructures or foreign counterparts on the other—especially coalitions cemented by exchanges

---

[16] The notion that a weapon's greatest effect may be to induce excessive defensive reactions on the other side is not unique to cyberspace. Richard Overy argued that the 8th Air Force's campaign in Europe, expensive as it was in manpower and money, nevertheless helped win World War II by persuading Germany to divert more resources that it could afford into building air defenses. Similar rationales were offered in support of the Reagan-era B-1 program and, later, the Strategic Defense Initiative (Richard Overy, *Why the Allies Won*, New York: Norton, 1995).

[17] DoD has institutionalized five "INFOCONs," alert conditions for the information world that parallel the five general defense conditions ("DEFCONs"), but they are not necessarily invoked at the same time. Although defense components have some leeway to react differently to the same infocon, the higher infocons can be associated with cyber lockdowns. (See Air Force Space Command, Instruction 33-107, *Information Operations Condition (INFOCON) System Procedures*, July 3, 2006.)

in cyberspace (which are bound to become more common).[18] Seams between alliances may be relatively soft targets in cyberspace. There are real reasons that such seams may permit infection. Rarely does a common security administration cover the target military and adjunct institutions (including other militaries). This makes unifying security architectures on the fly complex, even in the face of attacks that reveal inadequacies in existing architectures.[19] Cyberattacks can be crafted accordingly; an attack that crosses, strikes close to, or exploits the seam between institutions may result in finger-pointing and consequently less trust. Allies and allied organizations would face higher barriers to interaction with the target's network. From the inside looking out, external information sources could be cut back, and access to them from the inside would be restricted.[20]

Can a locked-down military work? Today, probably, yes. Militaries have traditionally been closed organizations, and closed militaries can still be deadly. No one expects nuclear establishments to be particularly accessible; ditto for al Qaeda. Even the U.S. military, with its vaunted degree of networking, cannot operate seamlessly in cyberspace

---

[18] See Libicki, 2007, pp. 128–136, 159–166.

[19] This process is so important precisely because attacks can slip through the seams between very different systems. Imagine that one system makes it very difficult for unauthorized users to gain access but gives its users a considerable of freedom within its boundaries. Next, consider a second system that erects fewer barriers to new users but restricts what users can do more tightly than the first system. Now, glue the two systems together in a network. At this point, a hacker may be able to gain access to the second system and acquire the privileges of a user. By being an authorized user of that system, it can also assert the privileges of being a user of the first system. This, in turn, allows it to access or corrupt that system using privileges it should never have had—and would not have had if it had had access only to the first or to the second system.

[20] This would not be tantamount to completely cutting off information, inasmuch as data can be conveyed through hard copy or, if such means can be trusted, via media exchange (e.g., thumb drives). But any added difficulties would reduce the flow of information and reduce the desire to work with allies and allied organizations vis-à-vis working entirely in house. The next generation of computer users may well find working with anything other than Web 2.0–like mechanisms for data exchange to be a hassle. Furthermore, if future exchanges involve interaction (e.g., let me show you how this operation would work by putting you in a simulation), they are nearly impossible to transfer without network-to-network interaction.

with allies—and it does not even come close to sharing operational information with those it works with daily. Yet it is hard to create high-tech organizations without networks that are at least partially open to the outside, if for no other reason than to permit logistics to interact forward with operations and backward with external support. Furthermore, opening access permits organizations to react to the unexpected by allowing individuals not normally in the designated information flow to offer fresh ideas and by facilitating the creation of ad hoc teams and relationships that allow new ways of working together.

So, coalitions still count—the more easily information flows among coalition members, the more easily they can coordinate action. The strengths of one partner can cover the gaps of another. The longer a conflict continues, the more important it is that warfighters learn—something that unconstrained access to information usually helps with. Conversely, organizations that wall themselves up, even if they have carefully fit their command and control to the logic of the systems they bring to war, will be unable to react with agility to the unexpected or reintegrate their operations in response to emerging opportunities and threats. Poor ideas will fester longer for their adherents having less opportunity to compare them with alternatives.

Ironically, another role for operational cyberwar would be the creation rather than the destruction of information—usually nearly worthless information. These days, destroying information has grown well nigh impossible: Information sources are multiplying, as are the means of accessing them. What operational cyberwar can do is to create channels through which operators can ramrod "information" to adversaries and thereby exacerbate their information overload. Trying to cope with this may induce particular organizational pathologies. The information need not be false, merely useless. Modest amounts of deception may persuade users to take it seriously enough to react. Granted, command-and-control infrastructures can react rationally to information overload, training users and designing systems to sift the wheat from the chaff more efficiently. But not everyone reacts to information overload calmly. If overload causes enough stress, the resulting reaction may be pathological. One such pathology is to overdele-gate, leaving the new information in the hands of lower organizational layers—which hinders integration. Another pathology is to overdiscount, not allowing new information to change fundamental cognitive categories—which hinders flexibility. Although such subtle strategies are potentially lucrative and persistent (not looking like attacks, they do not raise defenses), creating them requires detailed intelligence on the target's command-and-control systems and real-time intelligence on how the target is reacting.[21] Both needs underline the importance of intelligence to operational cyberwar.

Attacks on Civilian Targets

Disrupting or corrupting communications or transportation systems may help cripple military operations. Hitting civilian telephone switches may interfere with a military's command and control; even if military systems are separate, disabling other systems may hinder military mobilization. Nevertheless, many civilian targets of the sort that may be plausible targets for air raids (e.g., manufacturing plants) are not necessarily fruitful targets for cyberattacks. Since the cyberattacks can offer only temporary disruption, little may be gained by delaying, for a few days, a process that is more than a few weeks behind military operations in the supply chain.[22] Furthermore, using certain exploits against civilian targets reduces their efficacy against military targets.

A cyberattack on a civilian target may also be perceived as escalatory (although probably not as escalatory as a physical attack). Unfortunately, it may not always be obvious *when* a civilian target is being hurt. It is possible to know which military functions a network supports without knowing what civilian services the same network supplies.

---

[21] Libicki, 2007, Ch. 5, describes this logic more completely.

[22] This does not apply to a cyberattack whose aim is to corrupt a software or hardware component that goes into a military system rather than the system directly. The corruption may not be noticed until it is installed in the system and until the system runs through its processes—and that could be months or even years later.

Organizing for Operational Cyberwar

Operational cyberwar differs substantially from physical warfare in ways that should persuade commanders to rethink what they expect from cyberwarriors.

Take conventional warfare's emphases on readiness (the ability to go to war tomorrow) and sustainability (the ability to fight indefinitely). There is nothing wrong with these two concepts—except that they are unachievable in operational cyberwar to any practical extent.

Consider readiness: 10 USC distinguishes military services, which raise, equip, and organize forces, from component commands, which use forces to fight. The distinction presupposes that military units can be sent anywhere at any time to carry out missions. They do not have to be pretailored for their destination (weather gear and such aside). The parallel notion that cyberwarriors can be assigned to any target on the fly may not be entirely the case.[23] Computer systems are the same the world over, so hacking skills learned in one place should work just as well in another. But this tenet understates how much intelligence preparation is required for a successful attack. Success at operational cyberwar depends to a great extent on knowing where the target is vulnerable. Gauging the effect of a cyberattack has everything to do with the relationship between the affected information system and the human systems that rely on its information to make decisions. Both effects require intensive study of the specific target system. The more sophisticated the target, the more effort it will take to find usable vulnerabilities and understand the effects of exploiting them on adversary capabilities. Gaining such knowledge may not require physical presence—although the latter helps if tapping into people who know something about the target system would be useful—but it does require a high level of concentration on the target system that may be tantamount to mental presence.

Sustainability, as repeatedly noted, is not a useful expectation for operational cyberwar. Attacks exploit vulnerabilities. Exploited vul-

---

[23] In cyberspace, assignment need not mean physical movement to an operational center in the relevant area of responsibility. Nevertheless, colocation can foster person-to-person interaction of the sort that helps in coordinating military operations across warfare domains.

nerabilities, for that reason, are more likely to be fixed (if elucidated) or at least routed around (if not). The attacker needs new tricks, and such tricks are not always forthcoming, which is why using operational cyberwar first as a bolt from the blue and second as a special operation is so important.

Cyberwarfare *qua* warfare is soaked in intelligence. Preparation of the battlefield generally requires more effort—money, time, and people—than operating on the battlefield: Expenditure ratios of ten to one or 100 to one are quite plausible. The search for vulnerabilities is usually a search for specific vulnerabilities in specific systems that can be exploited in specific ways. Intelligence is also needed on network architecture, the relationships between various defense systems (e.g., what information from the target system feeds into which other systems?), and influence relationships (what information affects which types of decisions?).

Efforts to determine whether a cyberattack had its desired effect can be as important as efforts to generate such effects in the first place. BDA, the bane of operational cyberwar, is nevertheless essential to understanding what works and what does not (or, indeed, whether further effort is worthwhile). Relevant questions would cover the effects of the cyberattack on information flows (e.g., content, flow-rate), the mitigation techniques the target used in the wake of the attack, and the defensive measures the target took after the attack. Not surprisingly, when the U.S. government assigned an organization to provide the topmost defensive layer against attacks on military and civilian government systems in late 2007, it chose not a military service but an intelligence agency—NSA.

So, should operational cyberwar be carried out by intelligence agencies? Not necessarily. One can ignore the entire question about legal authorities. Even though warfare comes under the legal authority of 10 USC and intelligence under the legal authority of 50 USC, suffice it to say that military and intelligence lawyers can be quite creative in allowing both to work together.

Is culture the real issue? Intelligence operatives are oriented toward finding information about the adversary. Military operators are oriented toward reducing the adversary's ability to wage war, which in this context, generally means reducing the its ability to take advantage of information and communications. It is almost a cliché to note that the former want to keep communications going, the better to tap them, while the latter want to stop communications. Still, the difference should not be overestimated: NSA is, after all, run and staffed by military officers.

Institutional practices may be more relevant. Since intelligence is ongoing while operations are pending, intelligence operatives are likely to have accessed opposing systems well before military operators do. Operatives are more likely to be familiar with the target and may assume the initiative in decisions about what to attack. But with what priority? Intelligence gets its greatest credibility from its ability to ferret out secrets of state capabilities and intentions and convert them into knowledge for the national command authorities. The wartime variant of that is to divine the strategies of the enemy's decisionmaking infrastructure. Divining secrets of the adversary's SAMs lacks that kind of cachet and is likely to be a lower-status task.

Because cyberattacks are meant to cause system failure, attackers have to understand how opposing systems fail. Integrated air defense systems,[24] for instance, can fail by not seeing the target, seeing too many targets, failing to give or receive cuing information, not getting missiles to fire, firing missiles in directions that do not let them hit the target, or inappropriately emitting detectable energy. Military command-and-control systems have their own characteristic failure modes. Without a fundamental understanding of how such a system works, what can go wrong with it, and what enemy warfighters are likely to do (and do wrong) when it fails, a cyberattack on such a system would be no better than the proverbial shot in the dark. Those most likely to understand such failure modes—and thus those best placed to plan a military campaign that uses operational cyberwar— are likely to be those who understand how their own systems might fail. They are more likely to be military operators rather than intelligence operatives.

---

[24] The Air Force claims some confidence in being able to disable such systems via cyberattacks.

Yet those who prepare and conduct operational cyberwar will have to inject the intelligence operative's inclinations into the military ethos. These inclinations include seeking discrete rather than wholesale effects; the ability to wait patiently; an intuitive understanding that one is operating on the other guy's turf; a healthy wariness of deception, indirection, and concealment; and, yes, a willingness to abandon attack plans to keep intelligence instruments in place.

If intelligence and operations are twinned for cyberoperations, the long-standing issue about whether to exploit or destroy adversary nodes (or, in cyberspace, whether to exploit or expose taps into adversary nodes) might be resolved within the operational cyberwar outfit without bringing in higher-level decisionmakers. This introduces a final question, whose answer may seem counterintuitive to warriors who believe more is always better: How many people should be conducting operational cyberwar?[25]

Start with a general proposition: The number of successful attacks depends on how many exploits there are in hand. As it is, good general exploits are rare, while specific exploits can be used only against certain systems or when target systems are in a particular state (e.g., after a user accesses a particular Web site). The kinds and numbers of high-value targets the opponent has must also be factored in.

Although some might think "the more cyberwarriors the better," the available number of first-tier hackers may not be very large. Because exploits tend to depreciate rapidly after exposure—and thus often after use—these should be reserved for first-tier hackers. Too many second-tier hackers spoil the stew, and their activity might alert the adversary and provide it hints about the attacker's target sets and MOs and, worst, high-level exploits and implanted code. Even giving second-tier hackers low-level exploits to play with may be more likely to immunize rather than infect the target. Better tasks for these individuals would be mapping networks and dredging through user files.

---

[25] The discussion to follow is limited to computer network attack and does not include CNE; it also presupposes operations against a state's military information enterprise rather than against nonstate actors.

Conclusions

Living in an information age does not make operational cyberwar the be-all and end-all of military operations. If stretched that far, it could end up becoming nothing. Operators should also recognize that the best cyberattacks have a limited "shelf life" and should be used sparingly. If it is recognized for the rare and special thing that it is, operational cyberwar may have a few interesting roles to play.

Operational cyberwar is a support function, just as air warfare was throughout most of the 20th century. At this juncture, the world has never experienced operational cyberwar, except to a minute degree. The best guess is that operational cyberwar can be (1) a rapidly exploited, one-time bolt from the blue; (2) a carefully husbanded set of precisely aimed and timed effects; and (3) a wet blanket placed atop adversary ambitions to develop network-centric and coalition capabilities.

CHAPTER EIGHT
Cyberdefense

This monograph has strongly implied the importance of defending cyberspace thus far, largely because deterrence appears to be too problematic to offer much surcease from cyberattacks. Even DoD, which does have an offensive cyberwar mission, will likely spend *and need to spend* far more on defense than on offense—of which the ability to retaliate, hence deter, can only be part. A similar tilt is likely to characterize the U.S. Air Force as well, despite its global strike role.[1]

This chapter examines cyberdefense from the top down—first at the architecture, policy, and strategy level and only then at the operational level. The sequence is not dictated by the importance of the top; indeed, most of the effort to defend systems is inevitably the ambit of everyday sysadmins with the reinforcement of user vigilance. But for this reason, the nuts and bolts of cyberdefense are reasonably well understood and extensively written about, by RAND and others.[2]

---

[1] According to an item in *Air Force Times*,

> [24AF Commander, Major General] Lord said he envisions cyber operations at about 85 percent defensive—defending Air Force networks against attack and ensuring that networks aren't disrupted. The other 15 percent will be offensive actions against adversaries' cyber capabilities . . . . (Erik Holmes, "Donley Sets out Structure for Cyber Command," *Air Force Times*, February 26, 2009.)

[2] See, for instance, Philip S. Anton, Robert H. Anderson, Richard Mesic, and Michael Scheiern, *Finding and Fixing Vulnerabilities in Information Systems: The Vulnerability Assessment and Mitigation Methodology*, Santa Monica, Calif.: RAND Corporation MR-1601-DARPA, 2004; Robert H. Anderson, Phillip M. Feldman, Scott Gerwehr, Brian Houghton, Richard Mesic, John Pinder, Jeff Rothenberg, and James Chiesa, *Securing the U.S. Defense Information Infrastructure: A Proposed Approach,* Santa Monica, Calif.: RAND Corporation, MR-993-OSD/NSA/DARPA, 1999; Richard O. Hundley and Robert H. Anderson,

This discussion is limited to national defense systems, which are mostly but not exclusively military. Although many of the distinctions between military and nonmilitary systems are familiar, it may help to briefly reiterate three that are relevant to this chapter:

- Militaries have real enemies that wish to diminish them; other organizations have rivals but are more likely to be attacked for opportunistic or indirect reasons.
- Militaries generally do not have customers; thus, their systems have little need to be connected to the public to accomplish core functions (even if external connections are important in ways not always appreciated).
- Militaries are ordinarily on standby; they earn their keep by being prepared for extraordinary circumstances.

These circumstances do not necessarily make computer security more important or even better in defense systems than in nondefense systems. People from AT&T, the world's largest ISP, point to financial institutions as their most demanding and sophisticated customers. But bankers face a different set of cyberspace challenges than soldiers do and must seek their own solutions. So must the military.

The Goal of Cyberdefense

The first step in any cyberdefense policy is to determine the rules that such a policy is meant to enforce. This may be demonstrated by counterexample. According to the Webster Commission, much of the damage spy Robert Hanssen wreaked on the FBI was made possible because he was able to withdraw too much information from FBI com-

---

"Emerging Challenge: Security and Safety in Cyberspace," *IEEE Technology and Society Magazine*, Vol. 14, No. 4, Winter 1995–1996, pp. 19–28; Willis H. Ware, "Information Systems, Security, and Privacy," paper, Santa Monica, Calif.: RAND Corporation, P-6930, 1983; Willis H. Ware, "Perspectives on Trusted Computer Systems," paper, Santa Monica, Calif.: RAND Corporation, P-7478, 1988; and Willis H. Ware, *The Cyber Posture of the National Information Infrastructure,* Santa Monica, Calif.: RAND Corporation, MR-976-OSTP, 1998.

puters.[3] The main part of the problem was that he was accorded too many privileges; a minor part was that, although not a hacker, he knew a few tricks about how to get information surreptitiously. To complicate matters further, the September 11th attacks (several months after Hanssen was arrested) revealed the problems that are created when too little information circulates within the bureau.[4] The FBI responded to the Webster Commission report by making it much harder for internal hackers (FBI systems are air-gapped) to see and steal files they ought not. Unfortunately for the FBI, what it did not do was to assess comprehensively which types of agents were allowed to see what. The bureau did not touch the rules, only ensured better enforcement. Its reluctance to conduct a comprehensive assessment meant that the bureau generally defaulted to letting every special agent in charge decide.[5] It is debatable whether one or another operational architecture is correct, but it is not in question that, for the FBI at least, making no assessment at all yields imperfect results.

Before discussing *how* to defend, it helps to understand *why*. Of course, the goal of keeping all hostile activity outside the universe of defense systems is desirable and needs no further justification—but it has proven impossible so far, and the money to achieve it may not be cost-effectively spent.

Core military principles provide a guide to making cyberdefense achievable. If the purpose of having a military is to provide the abil-

---

[3] Commission for the Review of FBI Security Programs, *A Review of FBI Security Programs*, Washington, D.C.: U.S. Department of Justice, March 2002.

[4] The specific difficulty cited was that, although two groups of agents, one in Minneapolis covering Zacarias Moussaoui and one in Phoenix, both looked at flight schools and saw something anomalous, neither communicated its concern to the other. Romesh Ratnesar, Michael Weisskopf, Michael Duffy, Elaine Shannon, Maggie Sieger, and Bruce Crumley, "How The FBI Blew the Case," *Time*, June 3, 2002. However, two weeks after September 11th, the FBI asserted that it had known terrorists had been enrolled in flight schools but had had no information to indicate that the flight students had been planning suicide hijacking attacks. See Steve Fainaru and James V. Grimaldi, "FBI Knew Terrorists Were Using Flight Schools," *Washington Post*, September 23, 2001, p. A24.

[5] This, incidentally, had the secondary effect of creating a security architecture so complex that the FBI was famously unable to implement it in the now-failed successor to its mainframe-based and user-hostile Automated Case File system.

ity to exert military power (i.e., wage war and conduct other forms of defense), the purpose of cyberdefense is to preserve this ability in the face of attack. Full exclusion of all cyber mischief works. Yet if that is not attainable, such system qualities as robustness (which includes recoverability), integrity, and the ability to keep confidences are the true ends; exclusion is just one means to that end.

Robustness—the ability to extract as much military power from systems under stress as from systems free of stress—is no less important for information systems than it is for any other military system. It not necessarily a feature of any one system (although it could be) so much as it is of the military at the highest level of organization. It is the ability to absorb compromise and nevertheless operate almost as well as if nothing had happened. Ideally, the goal is to be able to make this claim: "Yes, perhaps you have broken into our systems, but notice that it has not slowed us down, increased our casualties, or decreased our ability to wreak damage on you." Recoverability is a key aspect of robustness—the ability to get some systems to cover for those that have been damaged while the compromised portions are being isolated, diagnosed, fixed, checked out, and returned to service.

The appearance of robustness is almost as important as robustness itself, if the goal is for military power to act as a general deterrent. The latent threat that another state may regard the military networks as DoD's Achilles heel, attack them, and start a war predicated on the success of a cyberattack is real. The threat can be neutralized if the attacker can be persuaded that its efforts would fail. Hence the urgent question: In that regard, what persuades?

Mearsheimer argues that nations are deterred from starting conventional war not by the fear of ending up as losers but by the prospect that the war will not be a cakewalk.[6] If so, a state that seeks to deter invaders is better off investing in the sort of deep, nested, robust defenses that make it difficult for the invader to win quickly. Investments that may offer the possibility of an eventual turnaround or offensive maneuvers in general would, everything else being equal, have less deterrent value.

---

[6]   Mearsheimer, 1983.

A simple analogy would suggest the same for cyberspace: Invest in robustness, protect core information-system capabilities, train to fight with degraded and suspect information systems, and emphasize the ability to reconstitute smartly following an attack. Nevertheless, such analogies can be too facile. The last thing a state wants is an enemy that thinks it is only a successful cyberattack away from paralyzing the state's ability to respond militarily. Thus, as important as it is to ensure that information systems are robust in and of themselves, it is more important to ensure that the military can offer a respectable defense in the face of a wide range of plausible cyberattacks. Whether that means an ultrahigh assurance of retaining 10 percent or 90 percent of its information capabilities depends on the relationship between military power and that least important 90 percent or 10 percent of the network.

Finally, as long as the contest is limited to cyberspace, cyberattacks alone cannot disarm the target's ability to conduct retaliatory cyberattacks. This makes the conflict a test of who can take punishment longer, in which case, the only way to lose quickly is to yield quickly. The ability to convincingly demonstrate a determination to ignore cyberattacks, regardless of their initial severity, may give the target the edge in such a standoff.

Note that robustness is not a result solely of cyberdefense per se; it also results from good engineering practice, albeit one that factors in not only malevolence but also error and accident. Indeed, the first principle of robustness is that systems (writ large, that is, systems of systems) be protected against the consequence of subsystem failure. One fundamental engineering principle, for instance, is that no set of software commands can cause a system to destroy itself inadvertently. Systems under stress should default to appropriate safe modes. Exactly what constitutes "safe" is context-dependent. For civilian machinery, a safe mode may be a controlled shutdown. Yet a design decision that a weapon should stop working if it gets unexpected commands may not be in the best interest of the warfighter, who depends on its working to keep the enemy at bay. Reverting to manual control may, for instance, be more appropriate. At the operational level, proper engineering must therefore determine the best default modes. For this reason, a book such as *Safeware*, despite its not mentioning hackers at all, is the kind of book that cyberdefenders need to read early on.[7]

Together with knowing what happens when systems fail, cyberdefenders should know how a system is likely to fail—its characteristic fault modes. The concept of characteristic does not necessarily mean most likely (although it could) but reflects where a system's components are most sensitive to failure or, conversely, what it has to get right to do its job. Loss-of-coolant accidents constitute characteristic failure modes for nuclear power plants, even if most nuclear plants that shut themselves off as a safety precaution have done so for other reasons. Outrunning one's logistics chain is a characteristic fault mode for armies. Inattentive airport security guards are a characteristic fault mode for airline security.

So, in preaching "know thyself," Sun Tzu and Plato were right, as much for cyberdefense as in philosophy. There is no substitute for understanding how a military system is likely to fail, if it is to be prevented from failing under a cyberattack. Further, cyberdefenders must vigorously seek out, if not embrace, failure—not only at the machine logic level but at the operational level (which is to say, the warfighting level). Such an assessment will be neither quick nor cheap (if it does not cost billions of dollars, it will probably have gotten short shrift), but its purpose is far broader than determining what cyberattackers can do. War entails destruction, and information infrastructures are at just as much risk as physical infrastructures. Military systems rely more heavily than commercial systems on RF links, and foes are likely to want to jam them. War induces stress even as it creates fog, and both lead to errors. There are, correspondingly, many ways for information systems to fail, and a thorough analysis of them is essential in making systems robust. Cyberdefenders will have to contribute to this process, but it is not their process alone.

After robustness, the next critical goal is *system integrity*: The system does what its operator wants it to do, and does not do what its operator does not want it to do. Integrity is also a function of the

---

7 Nancy Leveson, *Safeware, System Safety and Computers*, Reading, Mass.: Addison-Wesley, 1995. Anderson, 2008, does mention hackers and lots of them. It cannot be overpraised.

trust people have that the system will, in fact, do what it is supposed to. Although warfighters often have no choice but to trust their fighting machines, trust is more likely to be discretionary with information systems. An information system that is mistrusted but otherwise functional is just not very functional.

The last goal is *confidentiality*, the ability to keep secrets, not only those integral to the military's own operations but secrets others (such as the intelligence community) have entrusted to it. Although CNE is distinct from cyberattack, the two are strongly related. Preventing enemy access to information systems is an important component of both defenses. For better or worse, the military's information security architecture tends to be driven by older, paper-based paradigms in which security came more from protecting secrets than from ensuring robustness. Logistics information systems provide a case in point: These systems have less protection than operational systems because they store fewer secrets, but, as noted, they as critical as any system to the robustness of an operational military.

Architecture

U.S. military architecture is multilayered, consistent with the discussion on cores and peripheries in Chapter Two. It runs unclassified networks (the NIPRnet) that can access the Internet, support everyday communications of warfighters, and link DoD to its broader support community. It runs classified networks (SIPRnet) for command and control and the protection of sensitive information; such systems are air-gapped from the rest of the world. Finally, it runs various subnetworks at higher levels of classification to handle the most sensitive intelligence data. As far as has been publicly revealed, all the successful hacker attacks on military systems have targeted unclassified networks. In theory, nothing classified was revealed, but in practice not everyone has been fastidious about keeping classified information off such networks.

Three fundamental architectural challenges are (1) determining the division between unclassified and classified networks that best serves military purposes; (2) finding an appropriate level of protection for unclassified networks; and (3) ensuring that what are supposed to be isolated classified networks, are, in fact isolated.

One way to envision the proper distinction between NIPRnet and SIPRnet is to think of the former as the conduit for influence, liaison, and learning and the latter as a conduit for command and control (plus intelligence in yet more secure networks). Militaries prevail to the extent they learn rapidly,[8] and NIPRnet's security architecture should not impede learning. Despite the obvious security risks involved, the military must thus work together with coalition partners, form connections with the communities in which it works, and be in a position to exchange lessons learned with the broader national and domestic security communities. Ironically, one of the more insidious motives for cyberattack is to induce the digital equivalent of anaphylaxis: an immune reaction so powerful that it kills the body. A military that is frightened of interacting with others—and these days, unavoidably and increasingly through networks—is on its way to defeat. As one shrewd Army colonel once observed, "The first rule of information warfare is to not do it to yourself."

Somewhat different considerations apply to networks that can afford some information leakage but cannot afford a loss of integrity. Examples include logistics, medical support, environmental surveillance, and system monitoring. New security models may have to be sought for such systems, but these may be able to use filters or firewalls that are less fussy about what is sent out and that are much fussier about what is allowed in.

The key security challenge for DoD's classified, and presumably isolated, systems is ensuring that they are in fact isolated.[9] Potential breaches include physical access (e.g., unguarded doors), hardware-level access (e.g., inadequately monitored USB ports, unknown wire

---

[8] See, for instance, Overy, 1995, especially Chs. 4 and 10, and Eliot A. Cohen and John Gooch, *Military Misfortunes: The Anatomy of Failure in War*, New York: Random House, 1990, notably "Failure to Learn," pp. 59–94.

[9] Charles C. Mann, "The Mole in the Machine," *New York Times Magazine*, July 25, 1999, pp. 32–35.

connections, poor router configuration), component-level access (e.g., corrupted software or hardware inputs), RF access (e.g., systems that receive RF signals when they do not need to, RF links with no or weak encryption), and personnel access (e.g., untrustworthy users). Since classified networks can be stopped by faults in the machinery on which they depend, such a survey must also consider such hardware as routers that also handle unclassified traffic and even air-conditioning units connected to external networks. Because the likelihood of a breach is roughly proportional to the size of a network, the fact that SIPRnet has hundreds of thousands of access points suggests the low likelihood that it is completely air-gapped. That being so, every SIPRnet compartment has to be at least a little suspicious about what is coming at it from other parts of SIPRnet. Such utilities as anomaly-detection services, monitors, document circulation controls, internal firewalls, and the authorized ability to drop contaminated portions of the network cannot be entirely dispensed with.

Policy

If architecture is the framework for an organization's information system and if operations include the day-to-day techniques of defense, *policy* and *strategy* refer to the overall rules under which the information system runs. Most of the good principles of policy are familiar to the information security community and need not be repeated here. Simplification and correct task decomposition, however, merit especial note.

Insofar as complexity is the source of much vulnerability, simplification may be a friend of cyberdefense in the right context. The core problem is that, as computers become more powerful, what they "understand" exceeds what humans, who are evolving much more slowly, can understand. Yet it takes human understanding to detect that a system is acting "funny." This holds both for sysadmins (whose intuition is likely to be trained) and users (whose intuition is less assured). The simpler the system, the easier will be to explain its operations to commanders, improving their ability to integrate their own intuition about the reliability of a system into their other war plans. To simplify systems and what they hold, organizations concerned about information security should actively lean toward fewer features and options when the extras offer no more than occasional benefit. Thus, options in operating and application systems should default to disabled (hence, the value of putting open software in the hands of adept defense programmers). In electronic documents (broadly defined), extraneous material (e.g., loci for active code) should be filtered out. Network configurations should be as clean as they can be made. As long as zealotry is kept in check, the results ought to merit the hassle.

Correct task decomposition is the art of assigning cyberdefense responsibilities within an organization. The Air Force aphorism of centralized planning and decentralized execution fits cyberdefense. Architecture and the integrated analysis of system fault modes is a top-level job. Implementation of security rules has to be left with the same sysadmins who understand the networks they tend and whose job entails making things work in the face of hazards, induced and otherwise (most attacks on networks are actually attacks on the systems that sit on networks).

In emergencies, however, higher-ups may properly override local authorities when it comes to forensics and recovery. Top-level management of forensics is worthwhile because this expertise is concentrated and used only episodically. If larger attacks affect multiple systems in similar ways, top-down management of the recovery process can be more efficient and quicker (peer-to-peer information-sharing may also distribute useful tips). Recovery from a large attack also benefits from top-down management, particularly if it requires quarantine and triage. Quarantine may be required if networks host particularly virulent malware that can spread rapidly to others.[10] If the contagion is already widespread, quarantine may be required to ensure that networks that have yet to be cleaned do not reinfect clean networks. Triage is required when the cleanup resources are scarce and therefore must be allocated

---

[10] The assumption behind this sentence is that other systems are vulnerable. This is more likely to be so if the malware is new (e.g., a zero-day exploit) but is less likely if the malware preys on systems that lack patches when having patches is the norm.

to the more important networks (figuring out which ones are the most critical is yet another justification for self-knowledge). One particularly scarce resource may be people who are good at sensing whether files (including executable files and associated support code) have been tampered with. Against large attacks that consist of the repeated use of a few vectors, a top-down structure may be efficient at broadcasting the discovery and nature of such vectors throughout the affected population. Finally, to the extent that defense networks are like electrical networks, in that bringing everything back into service simultaneously may trigger outages, staged recovery—which also requires top-down control—is advisable.

Strategy

Most cyberattacks have a purpose. Discerning that purpose may offer a sense of what the attacker is trying to achieve—and allow the target to take steps to ensure that the attacker is pushed farther from its goal.[11] This can be dissuasive if it convinces the attacker that the cyberattack has backfired. An ill-considered cyberattack may even reveal the attacker's goals (for example that the attacker wishes the target to abandon a difficult international relationship with a third party, so that the attacker can swoop in and reestablish a relationship for its own ends).

More can be done to find ways to differentiate among motives for attack on military systems, such as determining whether a attack was a test, a feint, or the real thing. If the last, was it for immediate or somewhat later use? Tests, for instance, may be marked by sporadically high workloads for network operation centers correlated with unusual levels of outgoing packet traffic to unusual addresses. Those responding to such testing attacks should act as if they were being watched in the arena, e.g., by demonstrating flexibility and imagination (even if many of the adaptations are quietly ended later) but, at the same time, should quietly intensify their search for workarounds.

---

[11] This makes two assumptions: first, that the perceived strategy behind the attack is correct and not a ruse and, second, that pushing the other way makes sense in its own right.

Feints, because they have to be designed for repeatability, are likely to target the periphery rather than the core and are unlikely to use particularly sophisticated (hence damaging) exploits. Further evidence that an attack is a feint can be deduced from the state of the attacker's military readiness at the time. Indications that an immediate physical attack is being contemplated would include cyber attacks that are meant to disrupt rather than corrupt and attacks against logistics, mobilization, and deployment systems. The response, if it is not too late, should be to raise readiness and go to higher defense conditions, if appropriate. Demonstrating that the attacks had little effect may also help.

Corruption attacks, by contrast, point to later physical action because their effects are more difficult to discern and persist longer before being reversed. Indicators of these include code that fails integrity checks; hitherto unseen processes that monitor for outside signals; and, perhaps, unexplained errors in system tests. A good response would be to carry out demonstration exercises that emphasize the ability to operate without good systems and, at the same time, to intensify code checks and reboot systems found to have been corrupted.

Operations

Just as almost all policies for cyberdefense are familiar to the security community, so are almost all the operational techniques for cyberdefense. They need not be repeated here. Three classes of techniques merit note, though. Hardware signing (and hardware-based overrides) may be useful for an organization that (1) uses a great deal of hardware, much of which is distributed (e.g., sensors), and (2) can count on higher degrees of accountability than civilian organizations would be comfortable imposing. Deception, notably honeypots, honeynets, and decoys, deserves emphasis; these techniques not only divert attacks but permit characterization of the attacker's methods. Red teaming, valuable as it is for all institutions, is particularly relevant for military organizations, for fairly obvious reasons.

Hardware

The value and practicality of digital authentication, in general, and of hardware-based control, in particular, reflect the threat to defense systems and the expectation of strict accountability that can be expected of warfighters. Digital authentication can help distinguish real from fake messages and is also a useful countermeasure against malign insiders. Authentication ensures that, when someone or something receives a communication from another source, confidence is high that both the source and the communication are correct. If a rogue operator is sending communications that later prove problematic, blaming the operator (not some hacker) would be correct. Unfortunately, today's personal computers do not permit such confidence because their software is malleable and vulnerable to hackers (many integrated circuits can also be reprogrammed even after being embedded in a device). Thus, regaining confidence requires using hardware devices that cannot be reprogrammed. In some cases, these can be add-ons, e.g., stand-alone thumb-pads with USB connections.[12] In other cases, it may take persuading key vendors to install hardware devices (e.g., boot-up ROMs) that permit software installation only from specific vendors or from third-party vendors the original vendors have authenticated. Such ideas illustrate how the military can pursue a cyberdefense strategy specific to its unique threats and circumstances.

Deception

Military organizations necessarily have an interest in deception that goes beyond whatever help it provides in securing networks. By falsely portraying its networks and their contents, DoD can variously hope to persuade the attacker to direct its energies elsewhere; hope to misdirect the attacker's focus; and hope to give the attacker an exaggerated or

---

[12] Making such regimes work generally requires both two-factor authentication (e.g., a personal identification number) and personal accountability (warfighters are held responsible for losing devices, much as Israeli soldiers are held responsible when their weapons are lost). When DoD introduced an authentication system based on its Common Access Card, network intrusions fell by nearly half. See Bob Brewin, "CAC Use Nearly Halves DoD Network Intrusions, Croom Says," *Federal Computer Week*, January 25, 2007.

understated view of what it has been able to accomplish—not to mention foster a false impression of its physical capabilities.

Here, too, military and civilian networks differ. Hackers are interested in plumbing civilian networks primarily to inform their cyberattack efforts. State hackers, however, want insight on a military's physical capabilities to understand the strengths and weaknesses of those against whom they may have to fight—their doctrine, their command and control, their sensor coverage gaps, etc. The cyberdefender's job, in turn, is to create a distinct impression in the hacker's mind, in hopes that this impression will ultimately help misinform the hacker's leadership. The defenders' advantages should not be underestimated. Even if hackers "own" the networks they have successfully attacked, they lack the physical access to and the tacit knowledge of the system that resides in every operator's and administrator's head. They are, in essence, gazing at the vast cyberspace of their targets through a virtual peephole, and an agile defender can let them see only what is necessary to support the image that the defender wishes to project. To be sure, deception must be careful, lest one's own be deceived, but in this field, conscientious practice makes perfect.

The honeypot (as well as its cousin, the honeynet) is a well-known device for acquiring a sense of what cyberattackers are doing. If correctly configured, a honeypot ought to indicate the effects of the attack in pure form without the noise—the constantly altered files, the incessant exchange of packets over a network—that characterizes a normal computer in the hands of an active user.[13] With a properly configured honeypot, it may be easier to detect the presence of malware by noticing which files have been changed in unexpected ways or which processes the computer is attempting to carry out that it never before tried.[14] Honeypots can also be laden with processes or files that may be of particular interest to potential adversaries. Watching which files are copied or which processes are altered may also help elicit the goals

---

[13] Sophisticated attackers may also want to avoid honeypots to mask their techniques. If so, honeypots may have to carry out activities that make them look as though they are in active use but in ways that make it easier to factor out the activities generated by normal use.

[14] An unexpected change in process flow would be an indicator of an attack, for example.

of the cyberattacks and may illuminate the attacker's strategy in cyber-space.[15] The latter may provide useful hints about military posture or strategy. A bank, by contrast, has a variegated multiplicity of potential cyberthieves, making it less useful to try to understand each one.

Red Teaming

All organizations that face serious threats need to understand how well their networks stand up to attack. DoD has greater need than most, for several reasons. First, a serious attacker against DoD is unlikely to show its best moves until it makes a full-court press (e.g., at war). In contrast, such institutions as banks face many attackers who rarely need to wait years until external events have provided the right oppor-tunity. Second, militaries constantly undertake exercises, and a red-team exercise can easily fit into the patterns and structures available for other exercises. Third, militaries have access to intelligence feeds on particular enemies that make it possible to create particular attack modes (although limiting their exercises to modes that have already appeared would probably be a mistake). Fourth, since military facilities are generally better secured than civilian facilities, red teaming allows militaries to test the interaction between their cyber and physical pro-tection regimes.

Conclusions

Although DoD, in contrast to most other organizations, does not spend all its cyberattack dollars on defense, it still spends the vast majority of its resources there and deservedly so. Defense networks resemble noth-ing so much as civilian networks. For the most part, DoD uses the same hardware, software, and protocols. Thus, most of the tools and

---

[15] Typically, someone looking for an interesting file is likely to copy everything and sort through it all later, at leisure. These days, thieves are more likely to exfiltrate bytes at rela-tively slow rates for long periods to avoid being detected. If defenders can persuade thieves that the processes they put into computers can only take out so much at a time before their operation is shut down (and if thieves do not suspect they are being set up—a big if), they may develop and exhibit techniques to grab the files they want.

techniques DoD needs for defending its networks are the same as those for their civilian counterparts. This field has been well covered and need not be reviewed here.

Because of the military's unique characteristics, its cyberdefense strategy has to be framed accordingly. The most salient of these is the presence of potential adversaries with an interest in going to great lengths (e.g., selling corrupt devices for use in defense networks) to steal from, disrupt, or corrupt operations. Such foes also set great store in understanding military networks to understand the military itself. Military networks are also characterized by other differences: a user base used to strict accountability, the need to support emergency operations (i.e., war), and the high use of software-controlled devices (e.g., sensors, weapons). Thus, in many important ways, how such systems are defended will have important unique elements.

CHAPTER NINE

Tricky Terrain

When facing a threat that cannot be denied, a potential target can attempt to defend, disarm, or deter. As Figure 9.1 illustrates, the best mix of the three may depend on what mode of warfare is at issue.

In traditional land combat, for instance, the emphasis was on disarming the enemy in combat. Relying on defense generally did not work very well (the Maginot line being a prime example), and deterrence by threat of punishment generally required a disarmed or poorly

Figure 9.1
Where Various Forms of Combat May Fit in the Deter-Disarm-Defend Triangle

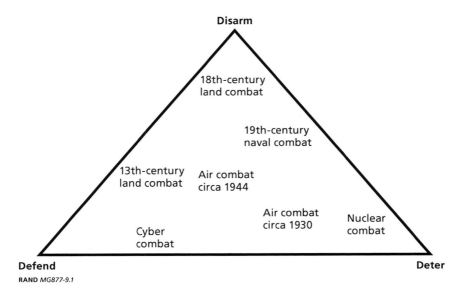

armed enemy (Sherman's march through Georgia, for example). In late medieval times, when castles were strong and artillery had yet to be introduced, the optimal point was closer to the defense apex of the triangle. In naval combat, the contest was historically over freedom of navigation; defense played a very small role (except at the tactical level, in terms of ship design). Yet there were vigorous debates between disarming (in terms of a "fleet in being") and deterrence by punishment (in terms of "commerce raiding"). Early observers of air warfare believed cities to be impossible to defend and were pessimistic about disarming invasion fleets, and so focused on deterrence.[1] As World War II commenced, populations evidenced a more stalwart attitude toward air attacks; disarming the Luftwaffe proved possible in the Battle of Britain, but ground-based air defense was often futile.[2] The optimal point moved toward the middle of the triangle. In the nuclear age, especially when expressed in terms of missiles, defense was nearly impossible, disarming ("counterforce") was a second-strike consideration, and so the primary emphasis was on deterrence ("countervalue").

What of war in cyberspace? Clearly, disarming is impossible, and no one seriously doubts that defense is necessary (lest joyhackers, criminals, and spies run riot through everyone's systems). The question is whether there is any role for deterrence. This investigation suggests that, in this medium, the best defense is not necessarily a good offense; it is usually a good defense. Granted, it is both extreme and unnecessary to foreswear deterrence (that is, to repeat, deterrence through punishment in kind) altogether, if only to have a credible response when some state openly challenges the United States in that realm. Otherwise, cyberdeterrence, as this discussion has demonstrated, is highly problematic. Before contemplating deterrence as its primary response to the threat of state-sponsored cyberattacks, the United States may first want to exhaust other approaches, such as diplomatic, economic,

---

[1]   Quester, 1986, pp. 82–104, argues that British fear of air attack—there was considerable panic after the Zeppelin raids in World War I—explained a great deal about Britain's capitulation in Munich.

[2]   Overy, 1995.

and prosecutorial means. At very least, the topic needs far more careful consideration than it has received to date.

Beyond deterrence lies combat in cyberspace, little of which is intuitive and less of which is straightforward. Ambiguity is ubiquitous. Every success in that medium relies on deception. Whoever declared the electron to be the "ultimate precision weapon" had or should have had his tongue planted firmly in his cheek.[3] In that medium, even the basic questions of journalism may lack answers: Who attacked? Where did they come from? What did they damage? How did they do it? When did the attack take place? Why did they bother? Cyberspace may be digital, but digital clarity is a property of high-definition television, not cyberwar.

To someone locked in a fight to the death, such ambiguities pose operational difficulties of the sort that apply to all weapons. Whether problematic weapons are useful depends on what it costs to employ them and what their likely effects are. By this criterion, cyber might look good. Cyberweapons come relatively cheap. Because a devastating cyberattack may facilitate or amplify physical operations and because an operational cyberwar capability is relatively inexpensive (especially if the Air Force can leverage investments in CNE), an offensive cyberwar capability is worth developing. An attacker can use them without worrying overly much about whether its own side suffers more than the target does, assuming a few reasonable precautions: Avoid creating self-replicating code (e.g., worms); make sure that the system you break is not something you or your friends actually use (without knowing it); etc. How much the military services should spend to build up an operational cyberwar capability is a classic how-much-is-enough issue. At this point, it appears that the operational corps should be small and elite, while the intelligence and planning cells should receive the most manpower and resources—a "measure twice, cut once" philosophy.

If combatants are not locked in a death struggle, the operational difficulties of cyberdeterrence; cyberwar; and, to a lesser extent, operational cyberwar risk becoming strategic difficulties—and all that fog-

---

[3] From testimony on June 24, 1996, by John M. Deutch, Director of the Central Intelligence Agency, before the Senate Permanent Committee on Investigations.

giness is not just about accuracy. If these elements are misused, the warrior risks fighting the wrong foe at the wrong time in the wrong place (e.g., in real space when cyberspace would have been less bloody) for the wrong reason.

Cyberdeterrence ought not to be confused with nuclear deterrence: Retaliation is so horrifying that none dares get close to testing it. Neither should it be confused with criminal deterrence: The law has a clear legitimacy advantage over the lawless. In an anarchic world, cyberattack and cyberretaliation might resemble the dynamic that governs rival urban gangs, mutually suspicious desert clans, or the Hatfields and McCoys—until someone escalates.

The dynamics of cyber confrontations also come perilously close to theater. Much depends on whether a confrontation plays out in the shadows, involving sysadmins, spooks, and presidents, or whether it plays out in the sunlight, where the whole world is watching. What may be a strategic minuet in the first case could well descend to farce in the second. What is actually happening can matter less than what people believe is happening.

Can the United States avoid cyberdeterrence and cyberwar altogether? Perhaps there is a foe so foolish as to attack the world's strongest military power by causing great annoyance to its society (perhaps by turning off everyone's lights, were it possible) and, in effect, asking: What are you going to do about it? The United States should probably be able to answer that query.

Whether any nation answers such a question in cyberspace, through less hostile means (e.g., prosecution, diplomacy), or through more violent means, however, would depend on a great many factors, not least of which is the confidence its leaders have in what really happened and why. Cyberretaliation—with all its difficulties—should not be the only response in the repertoire.

APPENDIX A

What Constitutes an Act of War in Cyberspace?

What constitutes an act of war may be defined one of three ways: universally, multilaterally, and unilaterally. A universal definition is one that every state accepts. The closest analog to "every state" is when the United Nations says that something is an act of war. The next-closest analog is if enough nations have signed a treaty that says as much. As far as cyberwar goes, no such United Nations dictum exists, and no treaty says as much. One might argue that a cyberattack is *like* something else that is clearly an act of war, but unless there is a global consensus that such an analogy is valid, cyberattack cannot be defined as an act of war.

The next way to define a cyberattack (with specified characteristics) as an act of war is to posit that a set of states has so defined cyberattack. NATO, the most obvious such organization, declared that the 2007 attack on Estonia did not merit invocation of the treaty's collective-defense clauses—not the same as denying that it was an act of war (problems with attribution may have been the key factor), but no ringing endorsement either. Had NATO declared that the attack *was* actionable, it might have served as a warning to potential attacking states, but whether they would have felt that this constituted a *legitimate* definition would be another matter. NATO would react to a cyberattack as it so declared, and the attacker would react to NATO's reaction as it deemed in its best interest. Legitimacy may play a role if the attacker did not believe that a cyberattack was as serious as a real attack and did not want NATO's reaction to serve as the last word on the subject.

Any state can unilaterally declare that a cyberattack (of certain characteristics) is an act of war, but acting on that declaration is another matter. Some states may find this declaration reasonable, legitimate, and actionable. Others may not. Potential attackers may or may not take such a declaration seriously. If the state responded to a cyberattack by retaliating, those skeptical of the claim might regard the response as illegitimate if it used a different modality from that of the attack itself. Whether such declarations are wise is a major subject of this essay. Nothing compels a state to treat cyberattack as an act of war.

Consider the following two vignettes. In the first, a rogue state, acting through a cutout (e.g., a phony engineering consulting firm), sends a manual to an electric power operator that persuades him to react to a thunderstorm by setting switches incorrectly. This error plunges the city into a week-long blackout and fries several hard-to-replace transformers. Dastardly perhaps, but this would probably not be regarded an act of war. In the second scenario, a rogue state employs a hacker to break into a computer system to change its instructions so that it reacts to the normal parameters of a thunderstorm (e.g., downed tree limbs severing power lines) by setting switches badly. The same effects result. If the first vignette is not an act of war, why would the second be?

Figure A.1 is one way to rank the seriousness of various types of cyber mischief. The gray line stands as a reasonable differentiation between cyber mischief that is actionable and cyber mischief that is not.

At the end of the day, the answer to whether a particular attack is an act of war comes down to this: Is it in your interest to declare it so?

Figure A.1
Ranking Various Forms of Harm in Cyberspace

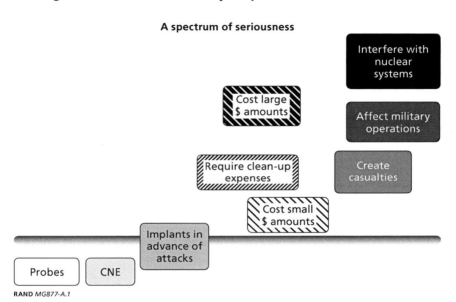

A spectrum of seriousness

Interfere with nuclear systems

Cost large $ amounts

Affect military operations

Require clean-up expenses

Create casualties

Cost small $ amounts

Implants in advance of attacks

Probes

CNE

RAND *MG877-A.1*

The Calculus of Explicit Versus Implicit Deterrence

Would the United States be better off having an explicit cyberdeterrence policy or maintaining its current implicit cyberdeterrence policy (that is, reserving a general right to retaliate at a time and in a manner of its choosing should it be deliberately hurt badly enough)?

At first blush, this does not seem like much of a question: Deterrence is in the mind of the potential attacker. What better way to persuade such attackers of the risks of aggression than by saying so in clear terms? Unfortunately, an explicit policy removes the purity of separating the easy cases ("we know who did it, and we can hit back") from the hard cases ("we are not sure about either") because others—attackers and third parties alike—will not be able to distinguish easily between unwillingness to retaliate and inability to know against whom or how to retaliate. Thus, a cyberattack that does not engender a response can undermine the credibility of the state with an explicit retaliation policy.

To explore this question systematically, we built a decision matrix, assigning probabilities and values to various outcomes, then did the sums to determine which posture offered the least expected pain. We did not generate a single solution using a canonical set of probabilities or outcome-values. Instead, our goal was to show what range of probabilities and outcome-values would likely make an explicit cyberdeterrence policy a good or a bad thing.

In this model, the relevant players are (1) the target state, which is also the potential retaliator and hence the decisionmaker in this exercise; (2) the (actual) attacker; and (3) the prime suspect, the state the target believes is most likely to have been the attacker. The target will retaliate only against the prime suspect, who may or may not be the actual attacker. There are also other states that have a nonzero likelihood of being the attacker (but need not fear retaliation as long as another state is perceived as the more likely attacker). Finally, there are third parties, whose opinion shapes the target's decision to retaliate. Third parties include everyone but the target state and the actual attacker; they represent more than just public opinion, since tomorrow's potential attackers (and even, in some circumstances, the prime suspect) may be among their ranks.

Apart from a few basic assumptions,[1] here are the core stipulations. If a state has an explicit rather than implicit deterrence policy,

- Other states will have a greater desire not to be *fingered* as cyber-attacker. Potential attackers could respond by attacking less or attacking *in less traceable ways.*
- Carrying out retaliation raises a state's credibility somewhat more—*somewhat* because the costs and benefits of retaliation depend in large part on what ensues, e.g., how deep an impression the retaliation made and whether the attacker counterretaliated.
- A state's reputation suffers much more if it does *not* retaliate; its bluff has been called. How much more depends on how likely it is that it was a state that carried out the attack (clearly a cyberattack by a nonstate actor does not call for retaliation against a state).

In this model, there are several basic outcomes, each with its own outcome-value. Each outcome-value has a label: **ouch** is the pain from being attacked; **wimpy** is the loss in face from letting the attacker get away with an attack; **risky** is the fear of igniting an extended conflict in cyberspace or, worse, against the attacker; and **oops** is the humilia-

---

[1] They are as follows: (1) Over the interval selected, there is one attack at most. (2) All attacks hurt the target to the same degree. (3) The target retaliates only against the prime suspect. (4) Nonstate attacks do not implicate states, and states do not take credit for such attacks. (5) The target is not at war with the attacking state. (6) The explicit declaration of an explicit deterrence policy is not per se costly or beneficial. All costs and benefits that accrue come from altering the likelihood of an attack and from the consequences of a response.

tion from being wrong, plus the risk of starting a fight with a justifiably aggrieved state. The outcome-value consists of the direct consequences of the attack plus the subsequent consequences that ensue from the decision to retaliate or not. There are four possibilities:

- No attack; no pain. Outcome-value equals zero.
- Attack; no retaliation. Outcome-value equals **ouch** (the direct consequences of the attack) plus some percentage of **wimpy**. The percentage of **wimpy** depends on how certain others are that the attack was the work of a state, which subsequently escaped punishment. As argued, the base value of **wimpy** is higher with an explicit deterrence policy.[2]
- Attack; correct retaliation. Outcome-value equals **ouch** plus **risky**. The risks have to do with the possibility that a general cyber-conflict or worse ensues.[3] As previously argued, the base value of **risky** ought to be somewhat mitigated if deterrence has been explicitly declared.
- Attack; incorrect retaliation. Outcome-value equals **ouch** plus **oops**. **Oops** is bad[4]—worse than **risky** or **wimpy**. Whether **risky** or **wimpy** is worse is an open question.

One might naively model the decision tree as in Figure B.1.

---

[2] It is one thing for a combatant to reject retaliation if it has never raised retaliation as a policy response (e.g., "oh, we do not retaliate in cyberspace, thanks for asking"). It is quite another for a combatant that has made retaliation a cornerstone of its announced cyberspace policy not to retaliate.

[3] There are other problems with retaliation. It may not work as well as touted. It may reveal vulnerabilities in the adversary's system that, when fixed, will render the systems harder against attack the next time around. Retaliation tends to legitimize cyberattacks in general and may excite the unwanted interest of third-party hackers.

[4] The retaliator is worse off in many ways. It has picked a fight with a state that it did not have to pick a fight with. That state, in turn, is probably more likely to counterretaliate if it was innocent than if it was guilty. Retaliation against the innocent cements a reputation as a bully. Finally, misplaced retaliation probably undermines deterrence. Although it shows the will to retaliate, it also shows an inability to discriminate between the innocent and the guilty. A state contemplating whether or not to carry out a cyberattack may well reason that, if it will face the retaliator's ire either way, why not go ahead?

Figure B.1
A Schematic of Cyberattack and Response

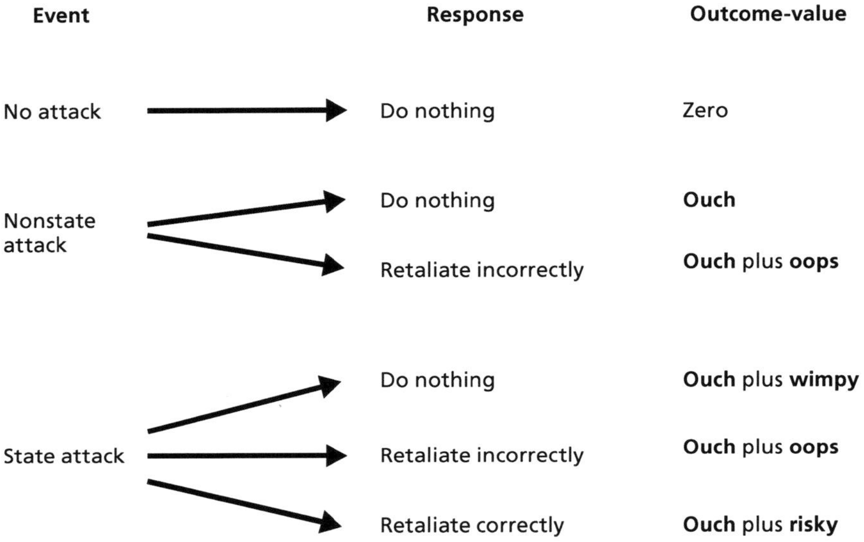

Event	Response	Outcome-value
No attack	Do nothing	Zero
Nonstate attack	Do nothing	**Ouch**
	Retaliate incorrectly	**Ouch** plus **oops**
State attack	Do nothing	**Ouch** plus **wimpy**
	Retaliate incorrectly	**Ouch** plus **oops**
	Retaliate correctly	**Ouch** plus **risky**

RAND *MG877-B.1*

But things are not so simple—if for no other reason than such a decision tree leaves no basis for the retaliator to determine whether it suffered a state or nonstate attack (and if the former, which state). Clearly, determining *ab initio* whether an attacker was a state or not, for instance, could eliminate from consideration the "retaliate incorrectly" option following a "nonstate attack" and limit consideration of the "do nothing" option following a "state attack." This would leave only the question of whether retaliation hit the right state.

Instead, it is important to recognize that the clarity of attack sources will vary greatly. In some cases, the attacker will clearly be a nonstate actor. In others, it may be unclear whether it was a state or a nonstate actor. With some, the fact that a state actor attacked is fairly clear, but *which* state is responsible is not. Then there are the attacks that could as plausibly be the work of a nonstate actor, the prime suspect, or yet another state. Thus, it is necessary to posit not the likelihood of an attack but the likelihood of an attack whose perpetrator is a matter of greater or lesser certainty.[5]

The argument about the variability of odds is not a mere quibble. If the source of every attack were obvious and if every attack appeared worthy of retaliation, an explicit retaliation policy would be clearly favored. This would inhibit attacks (reducing the odds of **ouch**), give more credit for retaliation, and avoid bluff-calling (which is particularly **wimpy** if one has declared one would retaliate).

The Decision Model

With that we can now establish our basic decision model.

The first thing we have to do is establish a menu of possibilities— that is, some likelihood for attacks of various levels of certainty.[6] To make things tractable, we grouped the initial event (or nonevent) into nine broad possibilities:

- no attack
- an attack that certainly came from a nonstate actor
- an attack from what was as likely to be the prime suspect as it was a nonstate actor
- an attack that might have been from the prime suspect and, if not, was from either another state or a nonstate actor
- an attack that might have been from the prime suspect and, if not, was from another state

---

5   Clarity of odds is almost always a good thing for the decisionmaker. If, in a given climate, rain takes place every other day in April, facing 50:50 odds may persuade an individual to carry an umbrella every day in April. A weather report that accurately indicated whether it definitely would or would not rain on a given day could reduce the need to carry that umbrella. Alas, such clarity is often missing with cyberattacks.

6   Technically speaking, we have to look at a probability distribution over a plane that bisects a cube made up of three axes: the likelihood that the prime suspect did it, the likelihood that another state did it, and the likelihood that a nonstate attacker did it. What makes it a two rather than a three-dimensional space is that the odds must sum to 100 percent.

- an attack that was probably carried out by the prime suspect but otherwise by a nonstate actor
- an attack that was probably carried out by the prime suspect but otherwise by another state
- an attack that was almost certainly carried out by the prime suspect but otherwise, by another state
- an attack certainly carried out by the prime suspect.

Each of these nine events has a certain probability, but that probability reflects both the world in which the target lives and how that world might respond to the target's deterrence policy.[7] As noted, an explicit deterrence policy ought to reduce the odds that a rational state will carry out an unambiguous attack on the target.

Before laying out the decision matrix, it helps to be clear about how the target would choose between an explicit and an implicit deterrence policy. It runs the following decision logic twice—once for an explicit policy and once for an implicit policy—and selects the one with the least expected pain:

- For each policy, the target assesses the likelihood of attacks of certain confidence levels (e.g., how likely is it that, in the next year, it will suffer cyberattack in which it is 75-percent certain that the prime suspect did it).
- It assigns the various outcome-values associated with **ouch**, **wimpy, risky**, and **oops**, as defined earlier.
- It calculates, for each type of cyberattack, whether its best course is retaliation or not (e.g., if the degree of confidence is 75 percent, then . . .).
- It calculates the expected outcome-value for the best course (which will include the **ouch** component).
- It sums the likelihood of a particular type of cyberattack multiplied by the expected outcome-value for the best course (e.g., retal-

---

[7] Presume that the odds these targets perceive are the true odds (as if the truth had been determined by flipping coins). Although being precise about the imprecision seems like an oxymoron, any other assumption would assume systematic bias on the target's part.

iate or not retaliate) in the wake of such an attack. By so doing, it finds out the overall expected outcome-value of its policy.

- It then compares the expected outcome-value from the explicit and implicit policies to determine which offers the least expected pain.

Table B.1 shows the decision matrix. Note that the two Odds columns remain to be filled in and that the terms *Wimpy X*, *Wimpy M*, *Risky X*, *Risky M*, and *Oops* remain to be converted into values.

Columns 1 and 2 define a set of possible attacks.

Column 3, the initial outcome-value, has two values: *Zero* if nothing happened or *Ouch* if a cyberattack took place.

Column 4, how the target responds, has two possibilities: *Do nothing* or *Retaliate* (against the prime suspect).

Columns 5 and 7 contain the odds (to be filled in) that that one of the nine possibilities ensues—with (column 5) or without (column 7) an explicit deterrence policy. The distribution of probabilities may be—in fact, ought to be—different (although they both sum to 100 percent). After all, if an explicit deterrence policy cannot make another state think twice about a cyberattack, what exactly was the point? Being explicit about retaliation may shift odds in other ways. Potential attackers may shape their attack in ways to reduce the certainty of attribution (e.g., so an attack where attribution was 90 percent likely might now be more ambiguous, pushing confidence levels down to 50 or 75 percent).[8] This is not a cost-free option for the attacker; if it were, it would be the default mode for most cyberattacks. It is also incompatible with some of the purposes for which a cyberattack may be contemplated (e.g., "Don't mess with me; I'm cyberbad."). Conversely, some other actor, knowing the United States is more likely to retaliate because it has an explicit deterrence policy and mindful that attribution is not always certain, might carry out an attack with the purpose of getting the target to retaliate against the prime suspect. Note

---

[8] Note that if the odds of a state attack being perceived as an attack by the prime suspect go down, the odds that of a nonstate attack or another state's attack being perceived as coming from the prime suspect have to go up if the total expectations of attack are to remain the same.

Table B.1
A Decision Matrix for Retaliation, Value Parameters

Odds of the Culprit Being		Initial Outcome Value	Action	Explicit Deterrence		Implicit Deterrence	
The Prime Suspect	Some State			Odds	Subsequent Outcome-Value	Odds	Subsequent Outcome-Value
No attack	No attack	Zero	Nothing		Zero		Zero
0	0	*Ouch*	Nothing		Zero		Zero
50	50	*Ouch*	Nothing		*WimpyX* × 0.5		*WimpyM* × 0.5
			Retaliate		*RiskyX* × 0.5 and *Oops* × 0.5		*RiskyM* × 0.5 and *Oops* × 0.5
50	75	*Ouch*	Nothing		*WimpyX* × 0.75		*WimpyM* × 0.75
			Retaliate		*RiskyX* × 0.5 and *Oops* × 0.5		*RiskyM* × 0.5 and *Oops* × 0.5
50	100	*Ouch*	Nothing		*WimpyX*		*WimpyM*
			Retaliate		*RiskyX* × 0.5 and *Oops* × 0.5		*RiskyM* × 0.5 and *Oops* × 0.5
75	75	*Ouch*	Nothing		*WimpyX* × 0.75		*WimpyM* × 0.75
			Retaliate		*RiskyX* × 0.75 and *Oops* × 0.25		*RiskyM* × 0.75 and *Oops* × 0.25
75	100	*Ouch*	Nothing		*WimpyX*		*WimpyM*
			Retaliate		*RiskyX* × 0.75 and *Oops* × 0.25		*RiskyM* × 0.75 and *Oops* × 0.25
90	100	*Ouch*	Nothing		*WimpyX*		*WimpyM*
			Retaliate		*RiskyX* × 0.9 and *Oops* × 0.1		*RiskyM* × 0.9 and *Oops* × 0.1
100	100	*Ouch*	Nothing		*WimpyX*		*WimpyM*
			Retaliate		*RiskyX*		*RiskyM*

that both effects—the prime suspect erasing fingerprints and another attacker adding the prime suspect's fingerprints—work to increase the likelihood of attacks that point weakly to the prime suspect. In the former case, it comes at the expense of obvious attacks; in the latter case, it comes at the expense of no attacks. Probably the only likelihood that will not change is the odds of an attack that is clearly not carried out by any state.

Columns 6 and 8 are the outcome-values for a combination of attack and response. Two aspects merit note. First, the outcome-value of **wimpy** has been replaced by *WimpyX* or *WimpyM*; similarly for **risky**. As noted, these values should be different: One would expect that doing nothing would hurt one's reputation more if one explicitly said one would react. The complex values for retaliation may also seem unusual, but they too make sense: If one retaliates against the prime suspect when the odds that the prime suspect did it are 50:50 (the lower of the two boldface cells), there is a 50-percent likelihood that the result will have a *RiskyX* outcome-value (the target was right) and a 50-percent likelihood that the result will have an **oops** outcome-value (the target was wrong). If one does not retaliate, the odds of being perceived as **wimpy** are proportional to the odds that some state did it.[9]

---

[9]   The assumption that the outcome-value of **wimpy** is proportional to the odds that some state did it is required to preserve linearity. This may strike some as forced. Third parties might plausibly excuse the target state for punishing neither state A nor state B if it cannot tell which is guilty, even though it is certain that one of the two is, in fact, guilty. One way to restore some realism to the assumption is to posit that half the onlookers believe state A certainly did it and that half believe that state B certainly did it, rather than to assume that all the onlookers believe the odds are 50:50 that either A or B did it. Given that assumption, the distribution of probabilities among specific states makes no difference to the perception that *some* state deserves retaliation.

Consider a cyberattack in the wake of which it is universally believed that the chances that state A did it are a; the chances that state B did it are b; and the chances that it was a nonstate actor are $1 - a - b$ (the total adding to 1). How **wimpy** will state A think the target is? If A did it, with a likelihood of a, the wimp factor is 1. If A did not do it, with a likelihood of $1 - a$, A will believe the odds that B did it are $b \div (1 - a)$—the odds that state B did it divided by the odds that someone other than state A did it (the odds that B did it and the odds that a nonstate actor did it must sum to 100 percent). The consolidated wimp factor in the eyes of state A is $a \times 1 + (1 - a) \times b / (1 - a) = a + b$. Similarly, the consolidated **wimpy** factor in the eyes of state B is the same: $b + a$. Finally, third parties have the same information the target does, that the chances that it was a state attack are $a + b$ and that the **wimpy**

To demonstrate the matrix, take the example cells highlighted in gray. The state has declared an explicit deterrence policy. A cyberattack takes place. Confidence that the prime suspect did it is 50 percent. If the prime suspect did not conduct the cyberattack, there is a further 50:50 likelihood that another state did. Otherwise, it was a nonstate actor. Is retaliation in order? The cost of rejecting this option includes a 75-percent chance of being seen as **wimpy**, which is particularly painful if the target has explicitly declared that it would not let attacks go unanswered. There is a 25-percent chance, however, that there is no one to retaliate against; if so, nothing is lost by holding back. Without retaliation, the outcome-value is 75 percent (50 percent plus 25 percent) times *Wimpy X*. The cost of retaliating is similar. With retaliation, the outcome-value is 50-percent chance of **risky** (e.g., the target will run the sort of risks that retaliation entails but will also bolster its word) and 50-percent chance of **oops**. When the least-cost option under such circumstances is determined, the target can calculate an expected cost (which includes the **ouch** factor) associated with that contingency: a cyberattack whose attribution runs 50:25:25 (prime suspect to other state to nonstate actor).

Intuitively, the choice to make deterrence explicit will not affect how target states will respond when certainty is very low (retaliation is thus never a good option here) or when certainty is nearly absolute (retaliation is thus always the better option here, if retaliation is ever going to be on the table). But it may well affect the certainty threshold for a response: The deeper the embarrassment of doing nothing, the more likely the target will act in ambiguous conditions.

Cross-multiplying the odds of each of nine contingencies by the expected pain associated with such a contingency under two deterrence postures—implicit and explicit—helps determine which posture leads to less pain, as demonstrated in the next section.

---

factor is thus *a + b*. Whether the target state weights its stature in the eyes of state A, state B, or the rest of the world is irrelevant because the **wimpy** factor for each is the same. Hence, what matters in calculating the **wimpy** factor is the likelihood that *some* state did it; the distribution of likelihood among states or the weight that the target places on its standing before various states is irrelevant.

Now, to Throw in Some Numbers

Table B.2 offers a sample set of numbers (shaded cells represent variables that are meant to be adjusted). Thanks to explicit deterrence, the likelihood of a cyberattack goes down from 85 to 80 percent—but the composition is quite different. Cross-multiplying the odds of a given type of attack by the odds that such an attack was carried out by either (1) the prime suspect, (2) another state, or (3) a nonstate actor reveals the overall odds of the target being attacked by a given source. Here, the odds of an attack by the prime suspect are reduced from 37 to 28.75 percent, but the odds of an attack by another state rise from 8 to 10 percent and the odds of an attack by a nonstate actor rise from 40 to 41.25 percent.[10]

Given this scenario—and recognizing that other numbers might produce a different net outcome, although perhaps similar patterns—a state with an explicit deterrence policy is actually worse off. With an implicit deterrence policy, the total pain is 122.35, of which 85 comes from the attacks themselves and 37.35 comes from the responses to the attacks. With an explicit deterrence policy, the total pain is 154.63, of which 80 comes from the attacks themselves and 74.63 (more than double) comes from the responses to the attacks.

Whereas the implicit deterrence policy would call for abstaining when the odds are 75 percent, the explicit deterrence policy would call for retaliating. The odds of having to retaliate go up only slightly (from 10 to 15 percent) because the odds of a cyberattack with which the likely suspect is certainly associated are eliminated. But the **oops** factor goes up from 0.5 percent (a 10-percent **oops** rate multiplied by 5 per-

---

[10] Incidentally, the identity of the prime suspect is not preordained. Another state may attack thinking the prime suspect will be blamed but may find itself labeled as the prime suspect. So, "the odds of an attack by another state" really equates to the odds of an attack by a state that cannot be identified with sufficient certainty to be labeled as the prime suspect. Also bear this irony in mind: Assume that the prime suspect does not attempt to hide its tracks better. Nevertheless, because nonstate actors and other states can shape their potential cyberattacks to look as though they come from the prime suspect, the target should be suspicious about an attack with particular characteristics and forensics really coming from the prime suspect. Ironically, therefore, if the assigned **wimpy** value does not change (i.e., *Wimpy X = Wimpy M*), declaring an explicit policy reduces the odds of retaliation.

Table B.2
A Decision Matrix for Retaliation, with Sample Numbers

Odds of the Culprit Being		Initial Outcome-Value	What We Do	Explicit Deterrence				Implicit Deterrence			
Prime Suspect	Some State			Odds	Subsequent Outcome-Value	Choice	Result	Odds	Subsequent Outcome-Value	Choice	Result
No attack	No attack	0	Nothing	20	0.000	Nothing	0.000	15	0.000	Nothing	0.00
0	0	1	Nothing	30	1.000	Nothing		30	0.500	Nothing	0.00
50	50	1	Nothing	10	1.500			10	0.750		
			Retaliate	15	2.650	Nothing	10.000		2.750	Nothing	5.00
50	75	1	Nothing	10	2.000			10	1.000		
			Retaliate		2.650	Nothing	22.500		2.750	Nothing	7.50
50	100	1	Nothing	10	2.000			5	1.000		
			Retaliate		2.650	Nothing	20.000		2.750	Nothing	5.00
75	75	1	Nothing	10	1.500			10	0.750		
			Retaliate		1.475	Nothing	14.750		1.625	Nothing	7.50
75	100	1	Nothing	5	2.000			10	1.000		
			Retaliate		1.475	Retaliate	7.375		1.625	Nothing	10.00

Table B.2—Continued

Odds of the Culprit Being				Explicit Deterrence				Implicit Deterrence			
Prime Suspect	Some State	Initial Outcome-Value	What We Do	Odds	Subsequent Outcome-Value	Choice	Result	Odds	Subsequent Outcome-Value	Choice	Result
90	100	1	Nothing	0	2.000				1.000		
		1	Retaliate		0.770	Retaliate	0.000	5	0.950	Retaliate	4.75
100	100	1	Nothing	0	2.000				1.000		
		1	Retaliate		0.300	Retaliate	0.000	5	0.500	Retaliate	2.50
						Total	154.630			Total	122.25

Values	
Ouch	1.0
Oops	5.0
RiskyM	0.5
RiskyX	0.3
WimpyM	1.0
WimpyX	2.0

cent odds of a cyberattack given 90-percent confidence that the prime suspect did it), to 3.75 percent (a 25-percent **oops** rate multiplied by a 15-percent odds of a cyberattack given only 75-percent confidence that the prime suspect did it). Since the outcome-value for **oops** is large (five times the pain of the cyberattack itself), this drives up the pain numbers for an explicit policy; so does the fact that the **wimpy** factor from not retaliating has doubled (again, part of the assumption).

What About Waiting?

Facing uncertainty in the wake of an attack, the target could wait to retaliate until further evidence has make the issue of who did it a little more clear.

By holding back on retaliation, the target may benefit from better information about who did it. For instance, what could be 60-percent odds that the prime suspect did it may, after investigation, resolve into a 25-percent likelihood that the prime suspect definitely did not do it (e.g., a nonstate actor did it) and an 80-percent likelihood (using some process of elimination) that the prime suspect did it.[11] In the first case, retaliation is clearly incorrect. In the second case, retaliation brings less fear of **oops**. Plugging in the numbers of the previous exercise (i.e., *oops* = 5, *riskyX* = 0.3, *wimpyX* = 2) and the 60-percent confidence (assuming the alternative to the prime suspect is that a nonstate actor did it), the cost of retaliating becomes 2.18 (5 × 40 percent + 0.3 × 60 percent) and the cost of abstaining is 1.2 (2 × 60 percent); abstaining is thus preferred. After waiting, the odds are 25 percent that the cost is 0 and 75 percent that the cost is either 1.24 with retaliation (5 × 20 percent + 0.3 × 80 percent) or 1.6 without (2 × 80 percent). Choosing retaliation as the lower-cost option gives a mixed score of 0.93, which is an improvement over the earlier 1.2 (it is a modest improvement because the activation threshold is 74.6 percent, which is not so far from 80 percent).

However, waiting has its costs (over and above the cost of investigation). Between the attack and the retaliation, the target may look weaker than it needs to look if it had had greater confidence in know-

---

[11] Seventy-five percent of 80 percent (100 percent less the 20 percent) is 60 percent.

ing who did it. Although resolution one way ("they were right to wait until they had more evidence") or the other ("finally, they struck back as they said they would") would erase some of the earlier impressions of fecklessness, it may not erase them all. The cyberattack will likely be off the news cycle, and the response will look more vindictive for not being so tightly linked in people's minds with the precipitating event. Not for nothing is "justice delayed" said to be "justice denied." If the purpose of the initial cyberattack was to test the target state, the attacker—observing the target's weak interim nonresponse—may press its advantage with more and perhaps different acts of aggression. Finally, there is no guarantee that waiting will necessarily improve the resolution of the uncertainty enough to make a difference.

Conclusions

The case for an explicit policy is stronger if it really does inhibit state behavior rather than force it to be sneaky and if it does not persuade others to attack, masquerading as the prime suspect. It helps if the target can afford the occasional mistake (**oops**), has a tolerance for risk if it retaliates (i.e., *RiskyX* is low), and is not viewed as particularly wimpy if it does not.

An implicit policy is preferred if a declaration cannot change state behavior in a helpful way (merely forcing attackers to be sneakier is not particularly helpful), if mistakes in retaliation are feared, if risk is to be avoided, and if there is a great deal more face to be lost if one declares in favor of but does not follow through with retaliation.

Invented numbers, even in otherwise good models, prove nothing, but they suggest that the case for an explicit deterrence policy is anything but obvious. They make plain what it takes for one or another policy options to be favored.

The Dim Prospects for Cyber Arms Control

Historically, arms control has always gone hand in hand with deterrence and crisis stability, but it would be difficult to be optimistic about its prospects in cyberspace. A good deal depends on what one means by arms control. If the model were to be something like the treaties signed between the United States–NATO and the Soviet Union–Warsaw Pact, which limited certain classes of weapons and banned others, there is little basis for hope.[1] If, instead, the goal were a framework of international agreements and norms that could raise the difficulty of certain types of cyberattacks, some progress can be made.

Why is it nearly impossible to limit or ban cyberweapons? First, although the purpose of "limiting" arms is to put an inventory-based lid on how much damage they can do in a crisis, such a consideration is irrelevant in a medium in which duplication is instantaneous.[2] Second, banning attack methods is akin to banishing "how-to" information, which is inherently impossible (like making advanced mathematics illegal). The same holds for banning knowledge about vulnerabilities.

Third, banning attack code is next to impossible. Such code has many legitimate purposes, not least of which is in building defenses

---

[1] See, for instance, Dorothy Denning, "Obstacles and Options for Cyber Arms Control," presented at *Arms Control in Cyberspace*, Heinrich Böll Foundation, Berlin, Germany, June 29–30, 2001.

[2] The one exception is a bot, for which numbers matter for determining attack effects. However, unless the attacking state keeps all its bots at home (in which case, such attacks would be easy to filter out), it has to subvert computers in other states—something which, in any case, is already illegal in the United States and in most of the developed world.

against attack from others. These others include individuals and nonstate actors, so the argument that one does not need defenses because offenses have been outlawed is unconvincing. In many, perhaps most cases, such attack code is useful for espionage, an activity that has yet to be banned by treaty. Furthermore, finding such code is a hopeless quest. The world's information storage capacity is immense; much of it is legitimately encrypted; and besides, bad code does not emit telltale odors. If an enforcement entity could search out, read, and decrypt the entire database of the world, it would doubtless find far more interesting material than malware. Exhuming digital information from everyone else's systems is hard enough when the authorities with arrest powers try it; it may be virtually impossible when outsiders try.

The only barely feasible approach is to ban the activity of writing attack code, then hope that the fear of being betrayed by an insider who goes running to international authorities prevents governments from organizing small groups of elite hackers from engaging in such nefarious activities. If the international community had the manpower and access to enforce such norms, it could probably enforce a great many other, and more immediately practical, norms (e.g., against corruption). Such a world does not exist.

More-probable (or at least less-fantastic) approaches may provide relief from the threat of cyberwar in treaties.[3] Such approaches would not prevent states from creating an offensive cyber capability and using it either overtly, covertly in the context of other aggression (e.g., a shooting war), or under circumstances in which they are prepared for the possibility of being caught. But it might inhibit states from using cutouts, paying freelancers, or tolerating mischief from within their borders. Measures would include stated norms that define cyberattacks and put nations on notice against such activities, treaties whose signatories pledge to assist other states' investigations of cyberattacks (even if the trail leads into sensitive organizations), and agreements not to use

---

[3] See Abraham D. Sofaer and Seymour E. Goodman, "A Proposal for an International Convention on Cyber Crime and Terrorism," Center for International Security and Cooperation, Stanford University, August 2000.

certain types of code (as measured by what it can do, rather than how it is written) when carrying out computer-network espionage. International support for more computer security research and development, as well as the implementation of secure versions of IPv6, would probably also help secure the global Internet.

References

Adee, Sally, "The Hunt for the Kill Switch," *Spectrum*, May 2008. As of August 25, 2008:
http://www.spectrum.ieee.org/may08/6171

Air Force Space Command Instruction 33-107, *Information Operations Condition (INFOCON) System Procedures*, July 3, 2006.

Anderson, Ross, *Security Engineering*, Indianapolis, Ind.: Wiley, 2008.

Anderson, Robert H., Phillip M. Feldman, Scott Gerwehr, Brian Houghton, Richard Mesic, John Pinder, Jeff Rothenberg, and James Chiesa, *Securing the U.S. Defense Information Infrastructure: A Proposed Approach,* Santa Monica, Calif.: RAND Corporation, MR-993-OSD/NSA/DARPA, 1999. As of November 24, 2008:
http://www.rand.org/pubs/monograph_reports/MR993/index.html

Anton, Philip S., Robert H. Anderson, Richard Mesic, and Michael Scheiern, *Finding and Fixing Vulnerabilities in Information Systems: The Vulnerability Assessment and Mitigation Methodology*, Santa Monica, Calif.: RAND Corporation MR-1601-DARPA, 2004. As of November 24, 2008:
http://www.rand.org/pubs/monograph_reports/MR1601/index.html

Bagchi, Indrani, "China Mounts Cyber Attacks on Indian Sites," *The Times of India*, May 5, 2008. As of August 25, 2008:
http://timesofindia.indiatimes.com/articleshow/3010288.cms

Barber, Bryan, "Cheese Worm: Pros and Cons of a 'Friendly' Worm," SANS Institute, 2001. As of August 25, 2008:
http://www2.sans.org/reading_room/whitepapers/malicious/31.php

Barboza, David, "Owner of Chinese Toy Factory Commits Suicide," *New York Times*, August 14, 2007.

Beagle, T. W., *Effects-Based Targeting: Another Empty Promise?* thesis, Maxwell AFB, Ala.: School of Advanced Airpower Studies, Air University, June 2000.

Becker, Gary S., "Crime and Punishment: An Economic Approach," *Journal of Political Economy*, Vol. 76, No. 2, 1968, pp. 169–217.

Benenson, Yaakov, Binyamin Gil, Uri Ben-Dor, Rivka Adar, and Ehud Shapiro, "An Autonomous Molecular Computer for Logical Control of Gene Expression," *Nature,* Vol. 429, No. 6990, May 27, 2004, p. 423. As of November 24, 2008: http://www.nature.com/nature/journal/v429/n6990/full/nature02551.html

Betts, Richard, *Surprise Attack,* Washington D.C.: Brookings Institution, 1982.

Bishop, Todd, "Should Microsoft Be Liable for Bugs?" *Seattle Post-Intelligencer,* September 12, 2003. As of June 12, 2009: http://seattlepi.nwsource.com/business/139286_msftliability12.html

Blank, Stephen, "Can Information Warfare Be Deterred?" *Defense Analysis,* Vol. 17, No. 2, 2001, pp. 121–138.

———, "Web War I: Is Europe's First Information War a New Kind of War?" *Comparative Strategy,* Vol. 27, No. 3, 2008, pp. 227–247.

Blau, John, "German Gov't PCs Hacked, China Offers to Investigate," *PC World,* August 27, 2007. As of November 24, 2008: http://www.washingtonpost.com/wp-dyn/content/article/2007/08/27/AR2007082700595.html

Boyes, Rogers, "China Accused of Hacking into Heart of Merkel Administration," *Times Online* (London), August 27, 2007. As of November 24, 2008: http://www.timesonline.co.uk/tol/news/world/europe/article2332130.ece

Brewin, Bob, "CAC Use Nearly Halves DoD Network Intrusions, Croom Says," *Federal Computer Week,* January 25, 2007.

Burgess, Christopher, "Nation States' Espionage and Counterespionage: An Overview of the 2007 Global Economic Espionage Landscape," *CSO Online,* April 21, 2008. As of November 24, 2008: http://www.csoonline.com/article/print/337713

Campbell, Matthew, "'Logic Bomb' Arms Race Panics Russians," *The Sunday Times,* November 29, 1998.

Cartwright, James E., Statement on the United States Strategic Command Before the House Armed Services Committee, March 21, 2007.

CERT Coordination Center and AusCERT, *Windows Intruder Detection Checklist,* Pittsburgh, Pa.: Carnegie Mellon University, 2006.

Chapman, Siobhan, "Worldwide PC Numbers to Hit 1 Billion in 2008, Forrester Says," *CIO.com,* June 11, 2007. As of April 27, 2009: http://www.cio.com/article/118454/

Chen, Thomas M., and Jean-Marc Robert, "The Evolution of Viruses and Worms," in William W. S. Chen, ed., *Statistical Methods in Computer Security,* Boca Raton, Fla.: CRC Press, 2004, pp. 265–285.

"Chernobyl Cover-Up a Catalyst for 'Glasnost,'" *MSNBC*, April 24, 2006. As of August 26, 2008:
http://www.msnbc.msn.com/id/12403612/

Cohen, Eliot A., and John Gooch, *Military Misfortunes: The Anatomy of Failure in War*, New York: Random House, 1990.

Commission for the Review of FBI Security Programs, *A Review of FBI Security Programs*, Washington, D.C.: U.S. Department of Justice, March 2002. As of November 24, 2008:
http://www.usdoj.gov/05publications/websterreport.pdf

The Conversation, dir. Francis Ford Coppola, American Zoetrope, the Directors Company, Paramount Pictures, and the Coppola Company, 1974.

CSIS Commission on Cybersecurity for the 44th Presidency, *Securing Cyberspace for the 44th Presidency*, Washington, D.C.: Center for Strategic and International Studies, December 2008.

Cusumano, Michael A., "Who Is Liable for Bugs and Security Flaws in Software? *Communications of the ACM*, Vol. 47, No. 3, March 2004, pp. 25–27.

Datong Li, "Xiamen: The Triumph of Public Will?" *openDemocracy*, January 16, 2008. As of November 24, 2008:
http://www.opendemocracy.net/article/china_inside/china_protests_or_politics

De Braeckeleer, Ludwig, "For Years US Eavesdroppers Could Read Encrypted Messages Without the Least Difficulty," *The Intelligence Daily*, December 29, 2007. As of November 24, 2008:
http://www.inteldaily.com/?c=169&a=4686

Denning, Dorothy, "Obstacles and Options for Cyber Arms Control," presented at *Arms Control in Cyberspace*, Heinrich Böll Foundation, Berlin, Germany, June 29–30, 2001.

Deutch, John M., Director of the Central Intelligence Agency, testimony before the Senate Permanent Committee on Investigations, June 24, 1996.

Erickson, Jon, *Hacking: the Art of Exploitation*, 2nd ed, San Francisco: No Starch Press, 2008.

Espiner, Tom, "US Reveals Plans to Hit Back at Cyber Threats, *ZDNet News*, April 2, 2008a. As of August 26, 2008:
http://news.zdnet.co.uk/security/0,1000000189,39378374,00.htm

———, "Georgia Accuses Russia of Coordinated Cyberattack," *CNET News*, August 11, 2008b. As of August 26, 2008:
http://news.cnet.com/8301-1009_3-10014150-83.html

"Estonia Fines Man for 'Cyber War,'" *BBC News*, January 25, 2008. As of August 26, 2008:
http://news.bbc.co.uk/2/hi/technology/7208511.stm

"Europe: A Cyber-Riot; Estonia and Russia," *The Economist* (London), Vol. 383, No. 8528, May 12, 2007, p. 42.

Evron, Gadi, "German Intelligence Caught Red-Handed in Computer Spying, Analysis," blog post, March 11, 2009. As of May 28, 2009: http://www.darkreading.com/blog/archives/2009/03/german_intellig.html

Fainaru, Steve, and James V. Grimaldi, "FBI Knew Terrorists Were Using Flight Schools," *Washington Post*, September 23, 2001, p. A24.

Federal Trade Commission, "ChoicePoint Settles Data Security Breach Charges; to Pay $10 Million in Civil Penalties, $5 Million for Consumer Redress," press release, January 28, 2006. As of April 29, 2009: http://www.ftc.gov/opa/2006/01/choicepoint.shtm

Fernandez, Manny, "Terrible Rumble, Then Chaos as Crane Fell," *New York Times*, March 16, 2008.

FRCC Event Analysis Team, "FRCC System Disturbance and Underfrequency Load Shedding Event Report February 26th, 2008 at 1:09 pm," final report, Tampa: Florida Reliability Coordinating Council, Inc., October 30, 2008. As of May 28, 2009: https://www.frcc.com/default.aspx

Gates, Robert, "Nuclear Weapons and Deterrence in the 21st Century," address to the Carnegie Endowment for International Peace: October 28, 2008.

Gelb, Leslie, *The Irony of Vietnam: The System Worked*, Washington, D.C.: Brookings Institution, 1979.

Gertz, Bill, and Rowan Scarborough, "Inside the Ring: NDU Hacked," *Washington Times*, January 12, 2007. As of September 4, 2009: http://washingtontimes.com/national/20070112-123024-8199r.htm

Gibson, William, *Neuromancer*, New York: Ace Books, 1984.

Gonsalves, Chris, "Security Quandary: Who's Liable?" *eWeek*, February 25, 2002.

Gorman, Siobhan, "Georgia States Computers Hit By Cyberattack," *Wall Street Journal*, August 12, 2008, p. A9.

Grow, Brian, Keith Epstein, and Chi-Chu Tschang, "The New E-spionage Threat," *BusinessWeek,* April 21, 2008, pp. 32–41.

Gunaratna, Rohan, *Inside Al Qaeda's Global Network of Terror*, New York: Columbia University Press, 2002.

Halderman, J. Alex, Seth D. Schoen, Nadia Heninger, William Clarkson, William Paul, Joseph A. Calandrino, Ariel J. Feldman, Jacob Appelbaum, and Edward W. Felten, "Lest We Remember: Cold Boot Attacks on Encryption Keys," *Proceedings of the 17th USENIX Security Symposium*, Berkeley, Calif.: USENIX Association, 2008, pp. 45–60. As of June 12, 2009: http://citp.princeton.edu/pub/coldboot.pdf

Harel, Amos, and Avi Issacharoff, *34 Days: Israel, Hezbollah, and the War in Lebanon*, Basingstoke, UK: Palgrave-MacMillan, 2008.

Harris, Shane, "China's Cyber Militia," *National Journal Magazine*, May 31, 2008.

————, "The Cybercrime Wave," *National Journal*, February 7, 2009, pp. 22–29.

Hart, Kim, "Longtime Battle Lines Are Recast in Russia and Georgia's Cyberwar," *Washington Post*, August 14, 2008, p. D01.

Hathaway, Melissa E., "Cyber Security: An Economic and National Security Crisis," *The Intelligencer: Journal of U.S. Intelligence Studies*, Vol. 16, No. 2, Fall 2008, pp. 31–36.

Herzog, Chaim, *The Arab-Israel Wars: War and Peace in the Middle East from the War of Independence Through Lebanon*, New York: Random House, 1982.

Hillbeck, Paul, "Argentine Judge Rules in Favor of Computer Hackers," *SiliconValley.com*, February 5, 2002.

Hines, Matt, "ChoicePoint Data Theft Widens to 145,000 People," *ZDNet News*, February 18, 2005. As of August 26, 2008:
http://news.zdnet.com/2100-1009_22-141381.html

The History of Computing Project, "Books on Hacking, Hackers and Hacker Ethics: An Annotated Bibliography," Web page, May 24, 2006. As of April 29, 2009:
http://www.thocp.net/reference/hacking/bibliography_hacking.htm

Holmes, Erik, "Donley Sets out Structure for Cyber Command," *Air Force Times*, February 26, 2009. As of April 29, 2009:
http://www.airforcetimes.com/news/2009/02/airforce_cyber_command_022509/

Howard, Michael, *The Franco-Prussian War*, London: Routledge, 1961.

Hundley, Richard O., and Robert H. Anderson, "Emerging Challenge: Security and Safety in Cyberspace," *IEEE Technology and Society Magazine*, Vol. 14, No. 4, Winter 1995–1996, pp. 19–28. Also available as RAND reprint RP-484. As of April 28, 2009:
http://www.rand.org/pubs/reprints/RP484/

Internal Revenue Service, "Suspicious e-Mails and Identity Theft," press release, June 13, 2008. As of August 21, 2008:
http://www.irs.gov/newsroom/article/0,,id=155682,00.html

International Telecommunications Union, "New ITU ICT Development Index Compares 154 Countries: Northern Europe Tops ICT Developments," press release, Geneva, March 2, 2009. As of June 1, 2009:
http://www.itu.int/newsroom/press_releases/2009/07.html

Israel Ministry of Foreign Affairs, "The Hamas Terror War Against Israel," Web site, August 3, 2008. As of August 26, 2008:
http://www.mfa.gov.il/MFA/Terrorism-+Obstacle+to+Peace/
Palestinian+terror+since+2000/Missile+fire+from+Gaza+on+Israeli+civilian+targets
+Aug+2007.htm

Jensen, Eric Talbot, "Computer Attacks on Critical National Infrastructure: A Use of Force Invoking the Right of Self-Defense," *Stanford Journal of International Law*, Vol. 38, 2002, pp. 207–240.

Johnston, David, and James Risen, "The Crash of Flight 587: The Investigation," *New York Times*, November 13, 2001, p. D9. As of June 12, 2009:
http://query.nytimes.com/gst/fullpage.html?res=9C03E2D61038F930A25752C1A
9679C8B63&sec=&spon=&pagewanted=2

Joint Publication 1-02, *DoD Dictionary of Military Terms*, Washington, D.C.: Joint Staff, Joint Doctrine Division, J-7, October 17, 2008.

Kahn, Herman, *On Escalation, Scenarios and Metaphors*, New York: Praeger, 1965.

Kash, Wyatt, "Spending for IT Security Gains Ground in 09 Budget," *Government Computer News*, February 7, 2008. As of August 25, 2008:
http://www.gcn.com/online/vol1_no1/45798-1.html

Kaufmann, William, "The Evolution of Deterrence 1945–1958," unpublished RAND research, 1958.

Kerber, Ross, "Grocer Hannaford Hit by Computer Breach," *The Boston Globe*, March 18, 2008. As of April 29, 2009:
http://www.boston.com/business/articles/2008/03/18/
grocer_hannaford_hit_by_computer_breach/

Korns, Stephen, "Botnets Outmaneuvered," *Armed Forces Journal*, January 2009, pp. 26–28, 38–39.

Krebs, Brian, "Virus Designed to Steal Windows Users' Data," *Washington Post*, June 25, 2004, p. A1. As of August 25, 2008:
http://www.washingtonpost.com/wp-dyn/articles/A6746-2004Jun25.html

Kuehl, Dan, "From Cyberspace to Cyberpower: Defining the Problem," in Franklin D. Kramer, Stuart H. Starr, and Larry Wentz, eds., *Cyberpower and National Security*, Washington D.C.: National Defense University Press, , 2009, pp. 24–42.

Lake, Eli, "McCain Backs Tougher Line Against Russia," *The Sun* (New York), March 27, 2008.

Lemonick, Michael D., "The Chernobyl Cover-Up," *Time*, November 13, 1989.

Lemos, Robert, "A Year Later, DDOS Attacks Still a Major Web Threat," *CNET News*, February 7, 2001. As of April 29, 2009:
http://news.cnet.com/2009-1001_3-252187.html

Leveson, Nancy G., *Safeware, System Safety and Computers*, Reading, Mass.: Addison-Wesley, 1995.

Leyden, John, Blu-Ray DRM Defeated, *The Register*, January 23, 2007. As of August 25, 2008:
http://www.theregister.co.uk/2007/01/23/blu-ray_drm_cracked/

Liang Qiao and Wang Xiangsui, *Unrestricted Warfare*, Beijing: PLA Literature and Arts Publishing House, 1999. As of June 12, 2009:
http://www.terrorism.com/documents/TRC-Analysis/unrestricted.pdf

Libicki, Martin C., *Conquest in Cyberspace*, Cambridge, U.K.: Cambridge University Press, 2007.

Lichbach, Mark Irving, "Deterrence or Escalation? The Puzzle of Aggregate Studies of Repression and Dissent," *Journal of Conflict Resolution*, Vol. 31, No. 2, 1987, pp. 266–297.

Loeb, Vernon, "Test of Strength," *Washington Post Magazine*, July 29, 2001, p. W08.

Mann, Charles C., "The Mole in the Machine," *New York Times Magazine,* July 25, 1999, pp. 32–35.

McClure, Stuart, Joel Scambray, and George Kurtz, *Hacking Exposed: Network Security Secrets and Solutions*, 5th ed., New York: McGraw-Hill Osborne Media, 2005.

McDougall, Walter, *The Heavens and the Earth: A Political History of the Space Age*, New York: Basic Books, Inc., 1985.

McMaster, J. R., *Dereliction of Duty: Lyndon Johnson, Robert McNamara, the Joint Chiefs of Staff, and the Lies That Led to Vietnam*, New York: Harper-Collins, 1997.

Mearsheimer, John J., *Conventional Deterrence*, Ithaca, N.Y.: Cornell University Press, 1983.

"Merkel's China Visit Marred by Hacking Allegations," *Spiegel Online International*, August 27, 2007. As of November 24, 2008:
http://www.spiegel.de/international/world/0,1518,502169,00.html

Meserve, Jeanne, "Sources: Staged Cyber Attack Reveals Vulnerability in Power Grid," *CNN.com*, September 26, 2007. As of April 29, 2009:
http://www.cnn.com/2007/US/09/26/power.at.risk/index.html

Miller, Jason, "Feds Take 'Cyber Pearl Harbor' Seriously," Homeland Security and Defense Business Council, May 29, 2007. As of April 23, 2009:
http://www.homelandcouncil.org/news.php?newsid=1091

MOCOM2020 Team, "41 Billion Mobile Phone Subscribers Worldwide," MOCOM2020 Web site, March 27, 2009. As of June 1, 2009:
http://www.mocom2020.com/2009/03/41-billion-mobile-phone-subscribers-worldwide/

Morgan, Forrest E., Karl P. Mueller, Evan S. Medeiros, Kevin L. Pollpeter, and Roger Cliff, *Dangerous Thresholds: Managing Escalation in the 21st Century*, Santa Monica, Calif.: RAND Corporation, MG-614-AF, 2008. As of April 28, 2009:
http://www.rand.org/pubs/monographs/MG614/

Morgan, Patrick, *Deterrence, a Conceptual Analysis*, Beverly Hills, Calif.: Sage Library of Social Research 40, 1977.

Mueller, John, *Retreat from Doomsday: The Obsolescence of Major War*, New York: Basic Books, 1989.

Mulvenon, James C., "The PLA and Information Warfare," in James C. Mulvenon and Richard H. Yang, eds., *The People's Liberation Army in the Information Age*, Santa Monica, Calif.: RAND Corporation, CF-145-CAPP/AF, 1998, pp. 175–186. As of April 28, 2009:
http://www.rand.org/pubs/conf_proceedings/CF145/

New Media Institute, "2007 CSI/FBI Computer Crime and Security Survey," Web page, September 14, 2007.

New York Independent System Operator, *Final Report on the August 14, 2003 Blackout*, Rensselaer, N.Y., February 2005.

Onley, Dawn S., "Army Urged to Step Up IT Security Focus," *Government Computer News*, Vol. 1, No. 1, September 2, 2004. As of August 25, 2008:
http://www.gcn.com/online/vol1_no1/27138-1.html

Overy, Richard, *Why the Allies Won*, New York: Norton, 1995.

"Protests Grow Over Chernobyl 'Cover-Up,'" *New Scientist*, No. 1688, October 28, 1989.

Public Law 105-304, 112 Stat. 2860, The Digital Millennium Copyright Act of 1998, October 28, 1998.

Quester, George, *Deterrence Before Hiroshima*, New Brunswick, N.J.: Transaction Books, 1986.

Randazzo, Marissa Reddy, Michelle Keeney, Eileen Kowalski, Dawn Cappelli, and Andrew Moore, *Insider Threat Study: Illicit Cyber Activity in the Banking and Finance Sector*, Pittsburgh, Pa.: CERT Coordination Center, Software Engineering Institute, Carnegie Mellon University, June 2005.

Ratnesar, Romesh, Michael Weisskopf, Michael Duffy, Elaine Shannon, Maggie Sieger, and Bruce Crumley, "How the FBI Blew the Case," *Time*, June 3, 2002. As of April 29, 2009:
http://www.time.com/time/magazine/article/0,9171,1002553,00.html

Rattray, Gregory J., *Strategic Warfare in Cyberspace*, Cambridge, Mass.: Massachusetts Institute of Technology, 2001.

Reed, Thomas C., *At the Abyss: An Insider's History of the Cold War*, San Francisco: Presidio Press, 2005.

Reilly, Michael, "How Long Before All-Out Cyberwar?" *New Scientist*, No. 2644, February 20, 2008, pp. 24–25.

Ridley, Kirstin, "Global Mobile Phone Use to Hit Record 3.25 Billion," Reuters, June 27, 2007. As of August 25, 2008:
http://www.reuters.com/article/technologyNews/idUSL2712199720070627?feedType=RSS

RSA Data Security, "RSA Code-Breaking Contest Again Won by Distributed.Net and Electronic Frontier Foundation (EFF)," press release, San Jose, Calif., January 19, 1999. As of April 29, 2009:
http://www.rsa.com/press_release.aspx?id=462

Russian Business Network, "Georgia CyberWarfare," blog posting, August 9, 2008. As of August 26, 2008:
http://rbnexploit.blogspot.com/2008/08/rbn-georgia-cyberwarfare.html

Sager, Ira, and Jay Greene, "Commentary: The Best Way to Make Software Secure: Liability," *Business Week*, March 18, 2002. As of August 26, 2008:
http://www.businessweek.com/magazine/content/02_11/b3774071.htm

Savage, Marcia, "Companies Still Not Reporting Attacks, FBI Director Says," *SearchSecurity.com*, February 15, 2006. As of April 29, 2009:
http://searchsecurity.techtarget.com/news/article/0,289142,sid14_gci1166845,00.html

Schelling, Thomas C., *Arms and Influence*, New Haven: Yale University Press, 1966.

Schiffman, Noah, "DARPA Attempting the Impossible: Self-Simulation for Defense Training," blog, *Network World*, June 6, 2008. As of April 29, 2009:
http://www.networkworld.com/community/node/28490

Schneier, Bruce, "Perspective: Internet Worms and Critical Infrastructure," *CNET News*, December 9, 2003. As of August 25, 2008:
http://news.cnet.com/2010-1001-5117862.html

———, "Information Security: How Liable Should Vendors Be?" *Computerworld*, October 28, 2004. As of August 26, 2008:
http://www.computerworld.com/securitytopics/security/story/0,,96948,00.html

Scott, William B., and David A. Fulghum, "Pentagon Mum About F-117 Loss," *Aviation Week and Space Technology*, Vol. 150, No. 14, April 5, 1999, p. 31.

Sevastopluo, Demetri, "Chinese Hacked into Pentagon," *FT.com*, September 3, 2007.

Shane, Scott, and Tom Rowman, "Rigging the Game," *Baltimore Sun*, December 10, 1995, p. 1A.

———, "Congress Has Tough Time Performing Watchdog Role," *Baltimore Sun*, December 15, 1995, p. 23.

Shawness, Lord, "Crime *Does* Pay Because We Do Not Back Up the Police," *New York Times Magazine*, June 13, 1965.

Shulsky, Abram N., *Deterrence Theory and Chinese Behavior,* Santa Monica, Calif.: RAND Corporation, MR-1161-AF, 2000. As of April 28, 2009:
http://www.rand.org/pubs/monograph_reports/MR1161/

Singh, Simon, *The Code Book: The Evolution of Secrecy from Mary Queen of Scots to Quantum Cryptography*, New York City: Random House, 1999.

Sofaer, Abraham D., and Seymour E. Goodman, "A Proposal for an International Convention on Cyber Crime and Terrorism," Stanford, Calif: Center for International Security and Cooperation, Stanford University, August 2000

Steiner, Peter, editorial cartoon, *The New Yorker,* Vol. 69, No. 20, July 5, 1993, p. 61.

Svensson, Peter, "Teen 'Unlocks' iPhone with Soldering Iron," *Mail & Guardian online*, August 26, 2007. As of August 26, 2008:
http://www.mg.co.za/article/2007-08-26-teen-unlocks-iphone-with-soldering-iron

Swartz, John, "Chinese Hackers Seek U.S. Access," *USATODAY.com*, March 12, 2007. As of May 8, 2009:
http://www.usatoday.com/tech/news/computersecurity/hacking/2007-03-11-chinese-hackers-us-defense_N.htm

Tang, Rose, "China Warns of Massive Hack Attacks," *CNN.com*, May 3, 2001. As of April 29, 2001:
http://archives.cnn.com/2001/WORLD/asiapcf/east/05/03/china.hack/index.html

Thomas, Pierre, "Teen Hacker Faces Federal Charges," *CNN.com,* March 18, 1998. As of August 25, 2008:
http://www.cnn.com/TECH/computing/9803/18/juvenile.hacker/index.html

Thornburgh, Nathan, "Inside the Chinese Hack Attack," *Time*, August 25, 2005a. As of August 25, 2008:
http://www.time.com/time/nation/article/0,8599,1098371,00.html

———, "The Invasion of the Chinese Cyberspies (And the Man Who Tried to Stop Them)," *Time,* August 29, 2005b. As of April 29, 2009:
http://www.time.com/time/magazine/article/0,9171,1098961,00.html

Traynor, Ian, "Russia Accused of Unleashing Cyberwar to Disable Estonia," *The Guardian*, May 17, 2007. As of August 26, 2008:
http://www.guardian.co.uk/world/2007/may/17/topstories3.russia

USC—*See* U.S. Code.

U.S. Code, Title 10, Armed Forces (10 USC).

———, Title 18, Crimes and Criminal Procedure (18 USC).

———, Title 50, War and National Defense (50 USC).

————, Title 50, Sec. 15, National Security (50 USC 15).

Vamosi, Robert, "Cyberattack in Estonia—What It Really Means," *CNET News*, May 29, 2007. As of April 29, 2009:
http://news.cnet.com/2008-7349_3-6186751.html

————, "Kids, not Russian Government, Attacking Georgia's Net, Says Researcher," *CNET News*, August 13, 2008. As of August 25, 2008:
http://news.cnet.com/8301-1009_3-10016152-83.html

Verton, Dan, "Blaster Worm Linked to Severity of Blackout," *Computerworld*, August 29, 2003. As of August 25, 2008:
http://www.computerworld.com/printthis/2003/0,4814,84510,00.html

Virginia Tech Department of Computer Science, "Hacking & Security Bibliography," Web page, Blacksburg, Va., May 2, 2002. As of April 29, 2009:
http://courses.cs.vt.edu/~cs3604/lib/Hacking/bibliography.html

Walker, Peter, "American Expats Caught Up in Indian Bomb Blast Inquiry," *Guardian.co.uk*, July 29, 2008. As of April 29, 2009:
http://www.guardian.co.uk/world/2008/jul/29/india.terrorism

Ware, Willis H., "Information Systems, Security, and Privacy," paper, Santa Monica, Calif.: RAND Corporation, P-6930, 1983. As of April 28, 2009:
http://www.rand.org/pubs/papers/P6930/

————, "Perspectives on Trusted Computer Systems," paper, Santa Monica, Calif.: RAND Corporation, P-7478, 1988. As of April 28, 2009:
http://www.rand.org/pubs/papers/P7478/

————, *The Cyber Posture of the National Information Infrastructure*, Santa Monica, Calif.: RAND Corporation, MR-976-OSTP, 1998. As of April 28, 2009:
http://www.rand.org/pubs/monograph_reports/MR976/

Wheeler, David A., and Gregory N. Larsen, "Techniques for Cyber Attack Attribution," paper, Alexandria, Va.: Institute for Defense Analyses, October 2003.

Whitaker, Mark, and John Wolcott, "Getting Rid of Kaddafi," *Newsweek,* April 28, 1986.

The White House, *Critical Infrastructure Protection*, Executive Order 13010,

Wilson, Clay, "Computer Attack and Cyberterrorism: Vulnerabilities and Policy Issues for Congress," Washington, D.C.: Congressional Research Service, April 1, 2005.

————, "Information Operations and Cyberwar: Capabilities and Related Policy Issues," Washington, D.C.: Congressional Research Service, September 14, 2006.

———, "Botnets, Cybercrime, and Cyberterrorism: Vulnerabilities and Policy Issues for Congress," Washington, D.C.: Congressional Research Service, January 29, 2008.

Wright, Lawrence, "The Spymaster," *The New Yorker,* January 21, 2008, pp. 121–122. As of May 8, 2009:
http://www.newyorker.com/reporting/2008/01/21/080121fa_fact_wright?currentPage=1

Yegulalp, S., "Review: Six Rootkit Detectors Protect Your System," *InformationWeek*, 2007. As of August 25, 2008:
http://www.informationweek.com/news/software/reviews/showArticle.jhtml?articleID=196901062

Zagare, Frank C., and D. Marc Kolgour, "Deterrence Theory and the Spiral Model Revisited," *Journal of Theoretical Politics,* Vol. 10, No. 1, 1998, pp. 59–87.

Zagorin, Adam, "Can KSM's Confession Be Believed?" *Time,* March 15, 2007. As of April 29, 2009:
http://www.time.com/time/nation/article/0,8599,1599423,00.html